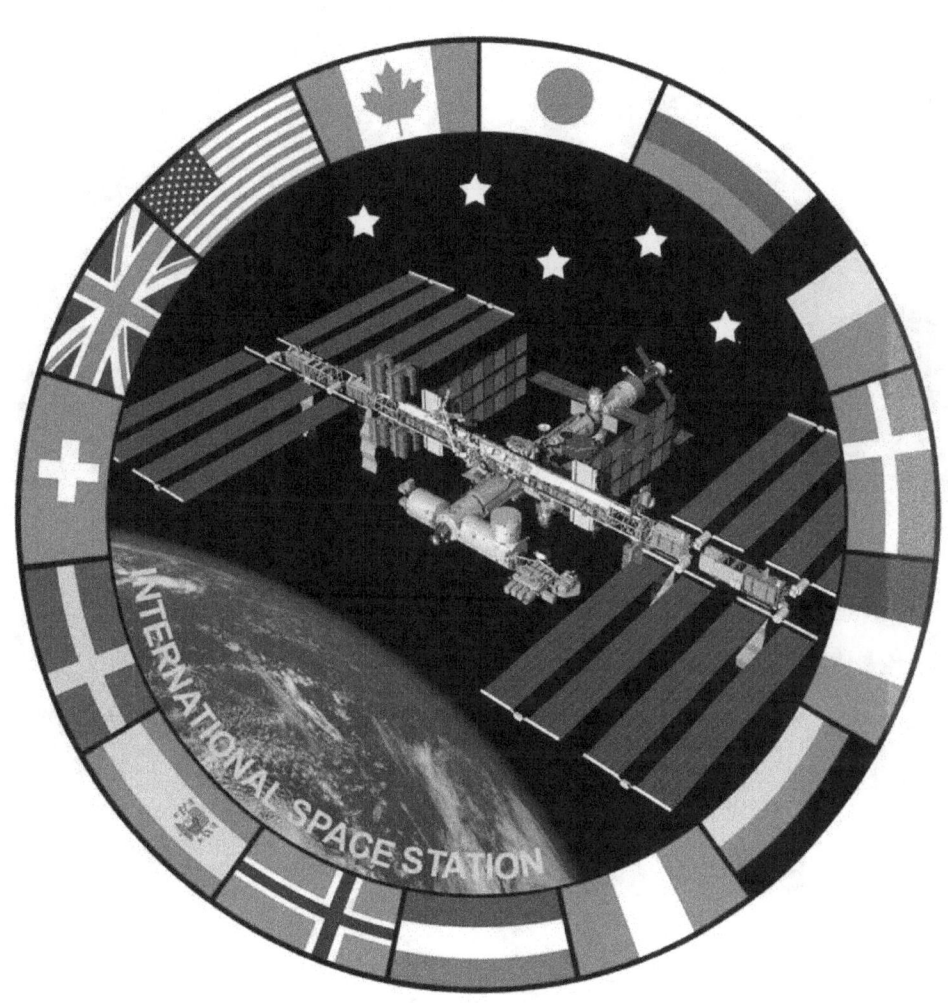

INTERNATIONAL SPACE STATION

The Era of International Space Station Utilization
Table of Contents

Astronaut Peggy Whitson looks at the plants grown in the Advanced Astroculture™ (ADVASC) green house.

Image: NASA ISS005E08001

Manfred Dietel
Charité Berlin, Germany

Berndt Feuerbacher
International Astronautical Federation, France

Vladimir Fortov
Joint Institute for High Temperature
Russian Academy of Sciences, Russia

David Hart
University of Calgary, Canada
Life Sciences Advisory Committee, Canadian Space Agency

Charles Kennel
Scripps Institution of Oceanography, USA
Space Studies Board, National Academy of Sciences, USA

Oleg Korablev
Space Research Institute
Russian Academy of Sciences, Russia

Chiaki Mukai
Space Biomedical Research Office
Japan Aerospace Exploration Agency, Japan

Akira Sawaoka
Daido University, Japan

Peter Suedfeld
University of British Columbia, Canada

Samuel C.C. Ting
European Organization for Nuclear Research (CERN), Switzerland
Massachusetts Institute of Technology, USA
Nobel Laureate in Physics

Peter Wolf
Observatoire de Paris, CNRS, LNE, Université Pierre et Marie Curie, France

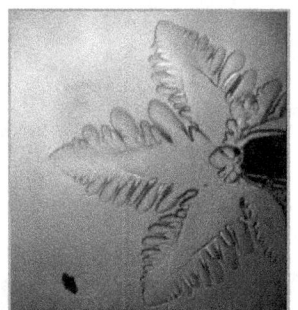

Physical Sciences

Fundamental Physics

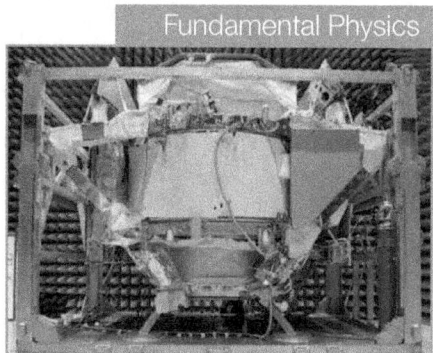

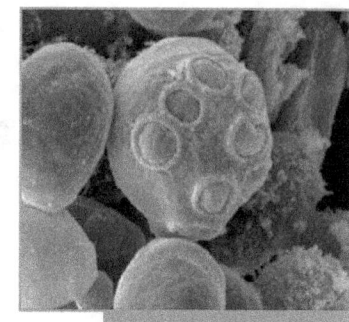

Life Sciences

Psychology and Space Exploration

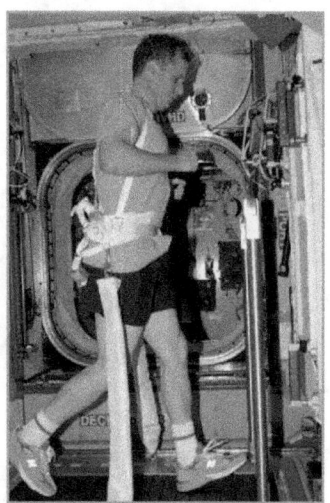

Earth and Space Observation

Human Health

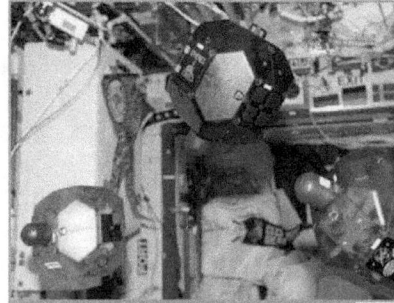

Exploration and Technology Development

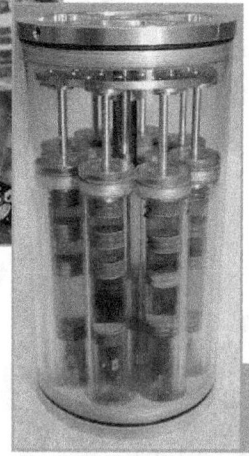

Commercial Development

Education

International Space Station research images.[1]

Executive Summary

Preamble

Human spaceflight is entering a new era. The assembly of the International Space Station (ISS) will be completed in 2010.[2] Supported by a full six-person crew for the first time, it is ready to put its full capabilities to work. While the ISS partners can be proud of having completed one of the most ambitious engineering projects ever conceived, the world at large also will judge the ISS by what is achieved in the utilization phase. In short, the full success of the ISS Program depends on the utilization achievements in the coming years. The people of countries participating in the ISS will expect no less.

For more than 15 years, the ISS partnership mastered financial and technical challenges, and weathered changes in national policies and governments. This mastery proves that nations can persist and achieve ambitious long-term goals that are very difficult. The ISS partnership is a model of what will be needed if an ambitious program of exploration beyond low-Earth orbit (LEO) is to move forward. The present partnership can be enriched by collaboration as we prepare for human exploration beyond Earth's neighborhood, to develop supporting technology, and to explore possible relationships with emerging space agencies. Aside from the ISS itself, the international partnership represents an invaluable achievement.

Extension of ISS operations to 2020 and beyond is crucial to maximize use of the ISS facilities. A longer operational phase provides opportunities for new participants who may never have thought of using the ISS. A commitment to extended operations enables programs with long-term objectives, and encourages institutions on Earth to support projects in space. Perhaps most importantly, extended utilization allows opportunities to explore the ISS as a research platform and to realize its full potential. Unique attributes of the ISS that enable research and development (R&D) never before achieved include:

(1) continuous access to microgravity[a] and defined partial gravity, enabling experiments with gravity as a controlled experimental variable;

(2) high vacuum and the conditions to create ultra-high vacuum,[b] enabling experiments that would be otherwise compromised by trace molecular species;

(3) continuous presence in the space environment, enabling long experiment runs and cumulative sets of experiments;

(4) significant power and instrument support services at a low-altitude (310-410 km) vantage point over 90% of the populated surface of the Earth, enabling use of the ISS as a platform for observations of Earth, Earth's atmosphere, and space processes;

(5) daily human support and transportation resources enabling testing, modification, and incremental development of R&D test beds, instruments, and research programs.

The benefits of ISS can be viewed from many different perspectives. As for other unique laboratories, long lead times to discovery can be associated with the many different disciplines that use the ISS. As scientists representing this broad array of disciplines, the ISS partner nations, and future utilization by all scientists worldwide, we

---

[a] 10^{-3} to 10^{-6} times Earth gravity (aka "microgravity").

[b] With additional pumping, the pressure behind a protective shield installed on the ISS perpendicular to the orbital velocity vector can be lowered to 10^{-12} (and even 10^{-14}) Pa.

met to discuss R&D on the ISS as the utilization era begins. We have captured the major disciplines, key questions, advantages of the ISS platform, and implications of ISS utilization for advancement of knowledge. The products of the discussion are a vision for the "Era of International Space Station Utilization," with supporting descriptions of the importance of the ISS to key R&D goals. We are certain that in the century to come, the full utilization of the ISS will be seen as having made transformative contributions to a number of scientific disciplines.

Benefits of the International Space Station

From experience gained in earlier human spaceflight programs (*Salyut*, Skylab, Space Shuttle, and *Mir*), and during the ISS assembly phase, the benefits of space-based R&D have already been demonstrated. These benefits include advancement of scientific knowledge, development of new technologies, development of Earth applications, continued growth in the international operations of a complex space endeavor, and the seeding of a global market for advanced space transportation. In addition to these benefits, the ISS offers an inspiring platform for inquiry-based educational activities that engage students in science, technology, engineering, and mathematics.

Scientific Knowledge: The ISS allows experiments to be conducted that cannot be done on Earth.

(1) The basic sciences of chemistry, biology, and physics can begin to treat gravity as an experimental variable over a broad range. The ISS provides continuous access to reduced gravity and microgravity for the first time. It is now possible to carry out systematic studies of processes masked by gravitational forces and subject to gravity-dependent phenomena,

NASA Astronaut Robert Behnken participates in an extravehicular activity (EVA) assembling the ISS in February 2010.
Image: NASA ISS022E062898

such as hydrostatic pressure gradient in fluids and the related sedimentation and buoyancy convection effects.

(2) Additional research areas can also be advanced on the ISS because of the capabilities and presence of such a capable platform in orbit. Although instruments and tests for such areas might have been possible on other platforms, they can benefit from the unique capabilities of the ISS (e.g., power, human tending) and increase the scientific yield.

(3) The use of the ISS creates the potential for research in social sciences (such as psychology, international relations, and economics) because of the unique conditions associated with living in space, conducting space operations, and forming and maintaining international alliances in science and technology.

The following is a non-exhaustive list of fields for which experiments and technology tests on the ISS offer potential for significant advancement:

- Astronomy and astrophysics
- Biology, exobiology, and biotechnology
- Health research/human adaptation and performance
- Earth sciences
- Fundamental physics
- Materials sciences
- Physics of fluids and combustion
- Solar system research and planetary science
- Space engineering and technology
- Space radiation research

New Technologies: The ISS, as a developed and full-service platform, allows an interactive process of testing without the need for a multitude of stand-alone missions and spacecraft.

(1) The ISS offers an unprecedented opportunity to test technologies for future exploration, such as spacecraft components, support systems, and mission operation scenarios. Wise use of the platform can provide space-proven technologies and reduce technical risk across other space endeavors and platforms.

(2) The configurability and human-tended capabilities provide unique opportunities to test research technologies, including instruments for fundamental physics, biology, medicine, and Earth observation.

Applications and Benefits: Scientific knowledge and new technologies derived from research on the ISS will benefit society through improving quality of life and fostering economic growth.

Return of Industrial Applications:
The production of benchmark samples (e.g., organic and inorganic crystals, functional nanomaterials, advanced alloys) obtained under controlled conditions and high-precision determination of material properties will contribute to industry for the optimization and development of ground applications. Such applications include: design of medical treatments and medicines; high-efficiency catalysts; and terrestrial production processes development. Specific industrial applications will also advance technologies for future space exploration, including space power and propulsion systems, fluids systems, and advanced materials.

A Model of International Cooperation:
Internationally managing a large remote research facility advances the peaceful use of space by all nations and provides the structure for future cooperation in the development of exploration missions beyond Earth's orbit. The operational structure for the ISS is now transforming from a focus on assembly to long-term utilization, and international operations are also transitioning to meet the research mission of a fully operational ISS.

Evolution of Space Transportation to the ISS:
To fully use the research potential of the ISS, it is critical to have robust transportation capabilities. The ISS has already led to the development of a variety of transportation vehicle systems to deliver cargo to orbit. Additionally, the ISS represents a stable destination over the coming one to two decades to allow the technical and operational evolution of crew and cargo transportation capabilities to and from LEO. This simultaneously supports ISS systems as well as utilization and opens new avenues for human exploration missions beyond LEO. Reliable transportation systems are essential for the delivery of samples and hardware to the ISS, and return of samples to Earth.

Recommendations

As research leaders, we have considered the benefits of the ISS to the partner nations and to the world. In addition, we note the benefits to science, engineering, and economic development, and we offer key recommendations to the agencies and governments that form the ISS partnership.

(1) Use of the ISS must be extended to 2020 and beyond to allow for broad and dynamic utilization. Implementation should be based on scientific standards, enable continuous evolution of significant new research objectives, and ensure that all the benefits of the global investment in the ISS can be realized.

(2) The ISS is a major stepping-stone for human space exploration. This platform provides invaluable long-term experience in operating a permanently occupied space complex with all the inherent logistic, medical, and technological challenges.

(3) The current ISS partnership should open access to the ISS for all nations to conduct research, technology development, and educational activities through a global call to the international scientific community. This broader community should be able to develop new ISS facilities and collaborate efficiently at an international level consistent with the global nature of scientific research. The first step in opening the platform to global research should be a declaration that the ISS is a unique Global Research Facility, open to all nations of the world for cooperative usage.

(4) To ensure extended use of the ISS through its entire functional lifetime, the partnership should develop a renewal process to reduce the uncertainty for the term of future use, enable an integrated research plan, and allow timely upgrades of facilities and instruments to best use its unique capabilities.

Scientific Disciplines and Potential

Gravity-dependent Processes in the Physical Sciences

By eliminating gravity or using gravity as a factor in experimental design, the ISS will allow physical scientists to better understand:

- fluid physics;

- the dynamics of interfaces, such as the line of contact between a liquid and a gas;

- the physical behavior of systems made up wholly or partially of particles;

- combustion processes in the absence of buoyant convection and their application to fire protection in spacecraft;

- the properties of molten materials and the processes during solidification.

Fluid Physics

Microgravity is particularly useful for the study of flow that is driven by surface tension, and for the study of the dynamics of fluids in structures such as foams and emulsions that are strongly affected by gravity. Studies in fluid physics also have ramifications to solidification of metals and semiconductors because the dynamics in the fluid state are important to the final properties of the solidified or crystallized materials. [3]

The Dynamics of Interfaces

In space, we can study the interfaces between liquids and gases without the interference of gravity. These studies have relevance to industry in the fields of energy production and food processing. [4]

The Physical Behavior of Particle Systems

The physical behavior of particle systems is relevant to many industrial processes. A better understanding of the behavior of dust particles also has implications for climatology, planetary formation, and planetary exploration (e.g., understanding the behavior of nanoparticles in lunar dust), and is also relevant to fundamental physics. The study of complex plasmas on the ISS has already yielded important advances. [5]

Combustion Processes in Space

In the absence of gravity and the accompanying absence of buoyant convection, combustion processes behave differently in space than on Earth. The ISS offers the facilities to explore and better understand flame propagation and other flammability issues for better fire protection in spacecraft. Research will lead to improved models of combustion in engines and of the ignition and propagation of fires in spacecraft.

A candle flame in Earth's gravity (left) and microgravity (right) showing the difference in the processes of combustion in microgravity.
Image: Glenn Research Center (NASA).

Material Melting and Solidification Processes

Studies on material melting and solidification processes, including advanced understanding of the physical properties of alloys and compounds using different solidification techniques and levitation, will be used to improve numerical models and to enhance the optimization of industrial metallurgical processes and the development of new advanced materials.[6]

Nickel-based superconducting dendritic crystals.
Image: Ames Laboratory, United States Department of Energy.

The interferograms shown below are from the Geoflow experiment, which is a model of Earth's crust and liquid core. Here, a viscous incompressible fluid (silicone oil) is used to understand fluid under different conditions. Applications include flow in the atmosphere and oceans, and movement of Earth's mantle on a global scale, as well as other astrophysical and geophysical problems. Results from Geoflow will also be useful for making improvements in a variety of engineering applications, such as spherical gyroscopes and bearings, centrifugal pumps, and high-performance heat exchangers.

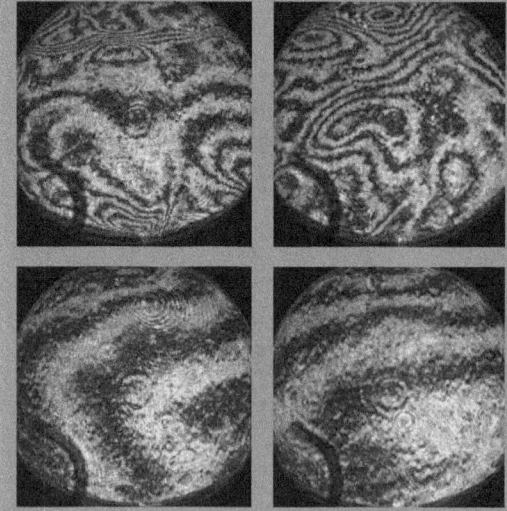

Image courtesy of Professor C. Egbers, BTU Cottbus.

The ISS represents the most capable laboratory in history for performing experiments in which gravity is controlled. The ISS offers the ability to perform repeated experiments over an extended period of time. The power and thermal capabilities of the ISS permit fluids and materials research plus combustion experiments that largely exceed any previously carried out in space. Scientists are not limited to a small subset of sample conditions, but can explore a broad range of experiment parameters by doing multiple sequential tests, just as they would do in their laboratories on Earth. The first Coarsening in Solid-Liquid Mixtures (CSLM) spaceflight experiment changed the way engineers use a classic theory of material design. Its findings have been incorporated into computer software for the design and manufacture of a wide range of products, from jet engines to suspension bridges.[7] Follow-on experiments on the ISS continue to expand our knowledge of materials science and its application for critical aerospace applications.

Fundamental Physics

The ISS can use the abundant power and serviceability inherent to ISS instruments to:

- search for dark matter, antimatter, and dark energy and to study energetic particles that cannot be studied on Earth

- test new technologies, such as atomic quantum sensors, that enable relativity tests to an unprecedented accuracy and can be used for the next generation of atomic clocks and other instruments on future missions

The Alpha Magnetic Spectrometer (AMS-02) is a state-of-the-art particle physics detector constructed, tested, and operated by an international team composed of 60 institutes from 16 countries and organized under United States Department of Energy sponsorship. The AMS-02 will use the unique environment of space to advance knowledge of the universe and lead to the understanding of the universe's origin by searching for antimatter and dark matter, and by measuring cosmic rays. AMS-02 is scheduled to be installed on the ISS in 2010.

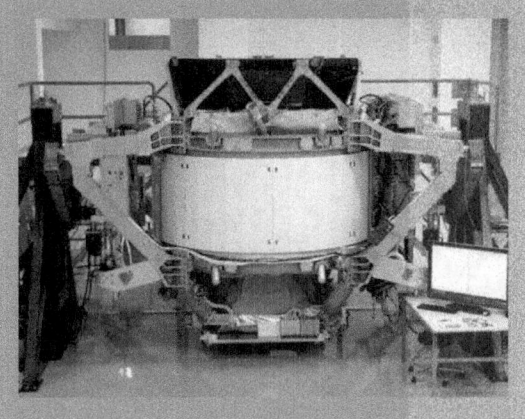

Image courtesy of the Massachusetts Institute of Technology, Cambridge, MA.

Atomic Clock Ensemble in Space (ACES), developed by the European Space Agency (ESA) and the Centre National d'Études Spatiales (CNES), is the most advanced experiment on atomic quantum sensors in space, planned for the ISS in 2013. The objective is to generate in space a stable and accurate time scale using laser-cooled atoms and to perform precision tests of Einstein's Theory of General Relativity through time comparisons with ground-based clocks located around the globe.

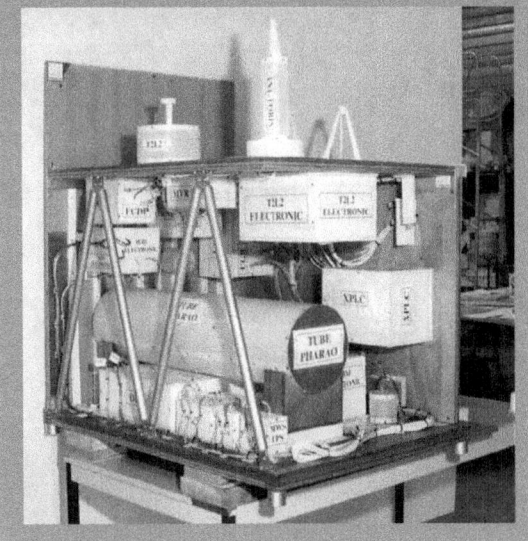

Image of early design mockup courtesy of ESA.

Particle Physics

The ISS offers a unique platform for future technology specific to fundamental physics such as observational tools (particle detectors, telescopes), atomic clocks,[8] electrostatic and atom interferometry inertial sensors, and optical and radio links for future clock comparisons and navigation. Testing of technology on a serviceable ISS is valuable before committing to dedicated experiments on other space platforms to probe the fundamental laws and constituents of nature.[9]

Gravity-dependent Processes in the Life Sciences

The ISS has scientific capabilities to provide a unique laboratory to investigate biological or life sciences without the constraint of gravity. Life scientists aim to answer the following questions:

- What is the role of gravity and genomic diversity in biological processes, and can such knowledge contribute to the solutions of biomedical problems that occur both on Earth and in space?

- What are the biological responses to multiple stressors?

- How can sustainable closed-loop biological life support systems be developed?

- Does life exist elsewhere in the universe, and, in that case, what is the origin, how did it evolve, how is it distributed, and what is the future of life in the universe?

- How do organic compounds (biomolecules and microorganisms) form in planetary atmospheres, and how are they transported?

Biology

Plants and animals have evolved and developed in gravity, and the role of this environment on the regulation of biological processes is only just beginning to be understood. Genetic diversity in some systems is obscured in the Earth environment; use of a microgravity environment should provide unique insights into such regulation. Previous microgravity studies observed increased virulence in microbes, pluripotency of stem cells, and tissue morphogenesis patterns. These early results indicate the potential for understanding gene expression and biological response to microgravity that cannot be studied on Earth. [10]

Results obtained from ISS research will have implications for understanding basic biological processes, understanding stress response, improving food supplies on Earth, and enhancing life-support capabilities for the exploration of space. In addition, better understanding of some of these biological processes (such as microbial virulence and the behavior of planktonic vs. biofilm forms of bacteria) could also have implications for astronaut health and provide crossover to translational activities between basic and more applied aspects of biological regulation important for improving human health on Earth. [11]

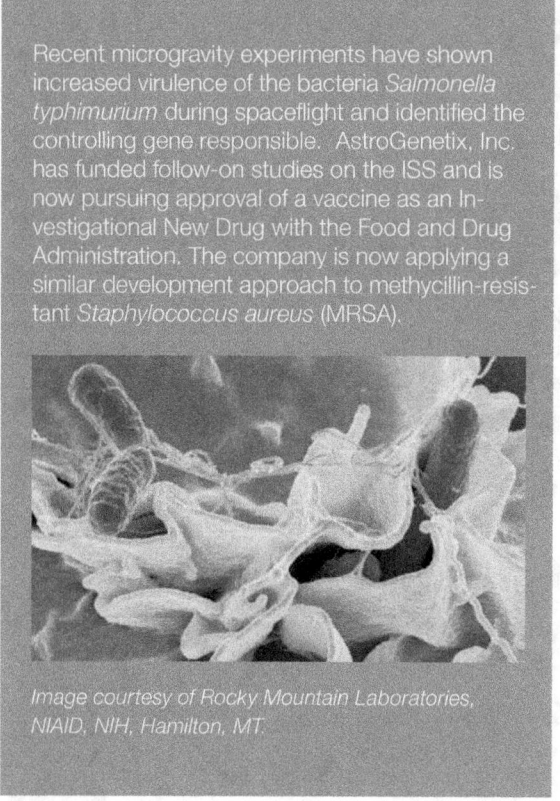

Recent microgravity experiments have shown increased virulence of the bacteria *Salmonella typhimurium* during spaceflight and identified the controlling gene responsible. AstroGenetix, Inc. has funded follow-on studies on the ISS and is now pursuing approval of a vaccine as an Investigational New Drug with the Food and Drug Administration. The company is now applying a similar development approach to methycillin-resistant *Staphylococcus aureus* (MRSA).

Image courtesy of Rocky Mountain Laboratories, NIAID, NIH, Hamilton, MT.

Astronauts are available to perform complex manipulations of culture and growth systems. Using the ISS laboratory facilities, multigenerational studies are readily completed on many different types of organisms. The ISS enables a surge forward in understanding the response of diverse life forms to gravity by offering repeated and frequent experiment opportunities. Results of one experiment can be applied to the next experiment in a matter of months.

Arabidopsis plants imaged in white light (left) and Green Fluorescent Protein excitation illumination (right).
Image courtesy of Robert Ferl, University of Florida, Gainsville, FL.

The ESA-sponsored Expose experiment contains several different biological specimens, such as the *Lichen Xanthoria elegans* seen below, that are exposed to the environment outside of the ISS.

Image courtesy of ESA: Columbus Mission Information Kit.[12]

Exobiology/Astrobiology

Exobiology (astrobiology) is the study of the origin, evolution, distribution, and future of life in the universe. The study of prebiotic chemistry and how the first building blocks of life were formed and the environment in which they formed remains a key exobiology question. Beyond prebiotic chemistry, exobiology is concerned with how organisms survive in space. This work encompasses the question of whether life can be transferred between planets. Organisms of scientific interest in the field include bacteria, fungi, multicellular organisms, communities, and biofilms. The ISS is the only platform allowing long-duration astrobiology experiments and return of samples for comprehensive analysis on Earth.

Human Health Research

The ISS provides an opportunity to examine human health in a way that cannot be done on Earth. ISS crews face physiological changes that can be experimentally reproduced and mitigated. These changes can serve as unique analogs for research on aspects of aging, trauma, and other physiological changes on Earth. Human health research on the ISS focused on improving the medical care of space explorers can answer the following questions:

- Are the effects of microgravity on the human body a good experimental model for diseases on Earth?

- What factors impair physical and cognitive performance, and what are the human physiological responses to multiple stressors?

- Can the differences in cell growth and differentiation observed in microgravity be used to address key questions of carcinogenesis and cancer treatment?

- Can one identify and validate optimal countermeasure strategies for the health of human space explorers based on physical, pharmacological, nutritional, and psychological interventions?

- What are the consequences (microbial environment, air composition, dust, etc.) of artificial life-support systems on human health and well-being?

- Can systems be developed that provide improved medical capabilities in human-rated spacecraft?

Physiological Processes and Human Health

In a microgravity environment, humans exhibit considerable individual variation in effects on a number of important physiological systems (bone and muscle loss; cardiovascular, immune, and neurological changes, to name a few) and

Investigations from the U.S. (Integrated Cardio-vascular), Canada (Vascular), Europe (Card), and Russia (Pneumocard) aim to determine the impact of long-duration spaceflight on the cardiovascular system, and will share data for the benefit of each other. Canadian Astronaut Dr. Robert Thirsk is shown exercising with the Advance Resistive Exercise Device (ARED) on the ISS.

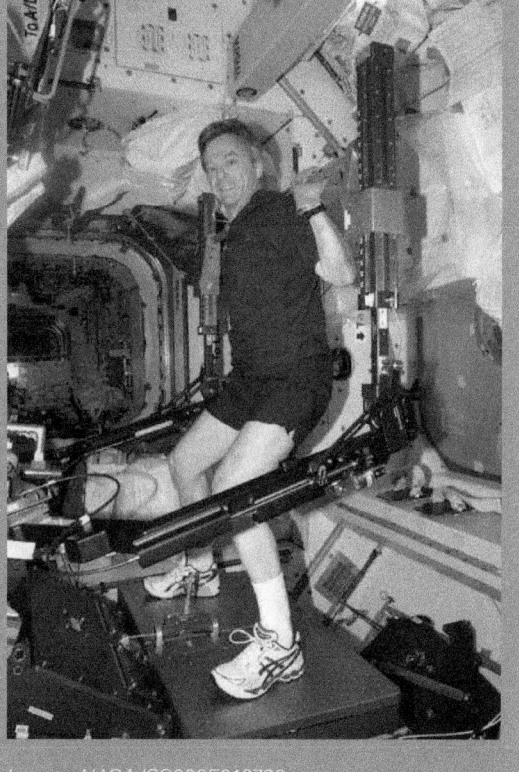

Image: NASA ISS020E010782

in their response to potential countermeasures. Additionally, in space, individuals are exposed to radiation risks, and genomic diversity could also influence the extent of this risk. Thus, research on human health in space offers potential insights that can benefit human health on Earth even as we gain understanding needed to develop effective countermeasures for safe, long-term space travel.

The ISS offers access to a unique, stable, long-term environment which can be used to assess the impact of microgravity on cells, model mammalian systems, and humans that cannot be duplicated on Earth. The current technologies available to

conduct studies in this space environment are excellent, and there is the potential to upgrade the capabilities as new technologies become available and build on the results obtained.

Use of the scientific capabilities of the ISS over the next 10 years and beyond should provide unique results regarding the impact of genomic diversity on human life, and generate further research directions to discover effective remedies in both microgravity and 1g environments.[13]

The training methods developed by the Advanced Diagnostic Ultrasound in Microgravity (ADUM) investigation have been incorporated by the American College of Surgeons Committee on Education into a computer-based program to teach ultrasound to surgeons, and have been used by the United States Olympic Committee to provide care during the Olympic Games. Below, ISS Commander and Science Officer Leroy Chiao performs an ADUM scan on the eye of Flight Engineer Salizhan Sharipov during Expedition 10.

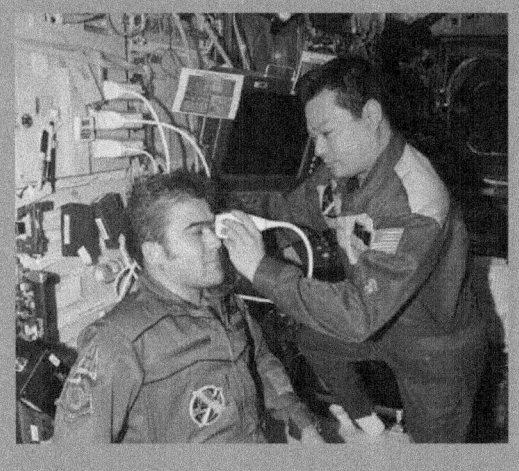

Image: NASA ISS010E18770

Cellular Differentiation and Applications in Cancer Research

According to the World Health Organization (WHO), cancer will be the most frequent cause of death in the western population by the year 2035.[14] This means that one of the major upcoming challenges in medicine is to find better treatment modalities for cancer and cancer-related diseases.

Directed cell growth, polarity of cells, differentiation, and cellular functions are partly influenced by gravity. This is assumed to be also true for the drug-cellular interactions that play the key role in drug-based cancer therapy. The microgravity environment offers the potential for a better understanding of malignant tumors and identification of new targets for cancer therapy that would not be detected under normal gravity conditions.[15]

Health Care Improvement to Benefit the Crew

Research on specific health care delivery systems focuses on crew health and mission success. The longer the mission, the more likely a medical emergency will occur, and response to the emergency will be limited by the absence of adequate diagnostic capabilities that are otherwise available on Earth.

Studies that use the ISS for advancing human health during long-duration flight include investigating autonomous health care capabilities, including telemedicine. Innovative medical equipment (both that developed on Earth and that developed for use in space) can be tested for viability on the ISS. Crew medical care on the ISS and future space missions is not so different from telemedicine needs in remote areas of Earth, on transoceanic flights, and in areas with limited availability of doctors, so there is great synergy between the disciplines.[16]

Psychology and Space Exploration

The ISS provides an opportunity to observe the interactions of multicultural crews working together in an extreme environment. This research seeks to answer the following questions:

- How can we assess and monitor health, psychological well-being, and interpersonal relationships in conditions of isolation?

- What are the factors governing the inter-individual variability in the response to spaceflight conditions?

- What are the roles of cultural and organizational factors on human performance during space missions?

Experience with previous spaceflights and in analogue and simulation environments clearly indicates that psychological and interpersonal problems are frequently related to adverse aspects of the physical and social setting.[17] Psychologists have begun to expand their vision from merely treating dysfunctions to enhancing emotional well-being (behavioral health) and optimizing performance, enjoyment, team cohesion, resilience, autonomy, and other favorable reactions (positive psychology). Future crews engaged in the exploration of planets will have to cope with unprecedented levels and kinds of stress. Enabling them to cope successfully will require systematic research to develop and test ways to enhance well-being as well as to intervene effectively against impairments in psychological functioning, crew interaction, and task performance.

The ISS is the optimal test bed for such research because it shares more characteristics with long-duration remote exploration than any other analogue or simulation environment.

Crewmembers from ISS Expedition 20 represent five nations and the five partners in building the International Space Station: Belgium (European Space Agency), Canada, Japan, Russia, and the United States.

Image: NASA ISS020E008898

Earth and Space Observations

Most remote observations of the Earth are conducted from dedicated orbiting platforms, and we expect this to continue. However, the ISS is a capable platform for instruments observing Earth and celestial bodies, and can support instrument servicing. There are important ways to leverage its presence in orbit to improve the return from instruments based on ISS and on autonomous satellites.

The following key questions in Earth and space sciences can be addressed now and in the future by testing and operating instruments on the ISS:

- What is the structure of the Earth system, and how is it changing? What is the structure and history of the universe? How has our solar system developed?

- How do spacecraft interact with the plasma environment of the Earth?

Technology Testing and Development

The ISS infrastructure[18] is the best platform to test experimental concepts, to promote compact and better performing instruments, to test new measurement methods and approaches, and to prepare innovative hardware for further utilization on automated platforms.

ISS facilities and infrastructure are equally valuable to test technologies related to Earth science, astronomy, astrophysics, observations of the solar system, and search for exosolar planets. Remote sensing instruments and space telescopes are large, expensive, and complicated facilities. For such payloads, ISS provides a unique opportunity of incremental assembly, module replacement, and regular service; e.g., liquid nitrogen and helium refill.

Plasma and Ionosphere Studies

In addition to remote observation of plasma and ionosphere, the ISS is also the largest artificial object within the plasma environment of the Earth. It gives a unique opportunity for in-situ measurements of plasma perturbed by spacecrafts and thrusters of various kinds, and for active experiments.[19]

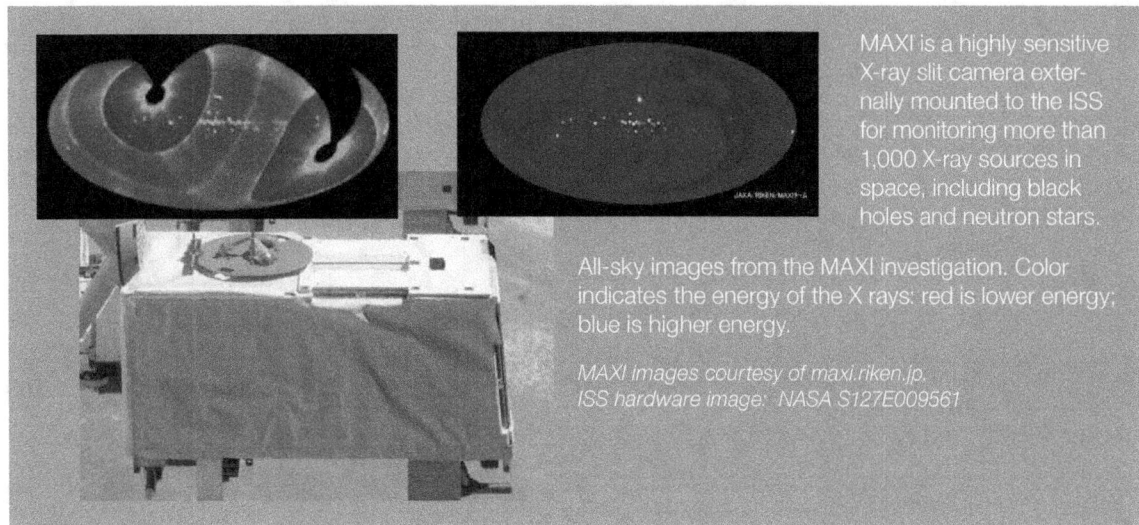

MAXI is a highly sensitive X-ray slit camera externally mounted to the ISS for monitoring more than 1,000 X-ray sources in space, including black holes and neutron stars.

All-sky images from the MAXI investigation. Color indicates the energy of the X rays: red is lower energy; blue is higher energy.

MAXI images courtesy of maxi.riken.jp.
ISS hardware image: NASA S127E009561

Exploration and Technology Development

The ISS provides an opportunity to test technologies needed for future human exploration of space. This research seeks to answer the following questions:

- How can the ISS be used to advance the state of the art in robotic technologies to extend both robotic and human presence further into the solar system?

- Can the development of novel materials on the ISS enable new strategies for the construction of space vehicles and space platforms?

- Can life-support systems provide increasing independence of human-rated spacecraft from resupply from Earth?

- How can the ISS, as the only space platform that allows long-duration human habitation, best be used to propel technology in support of human exploration and space technology that can be adapted for benefits on Earth?

Robonaut2, the next generation of dexterous humanoid robots, was designed through a joint venture between NASA and General Motors.

Image: NASA JSC2009-E-155295

To expand the human presence beyond LEO, significant advances in technology will be needed. The major challenges of human spaceflight beyond the near-Earth orbit include: the severe radiation environment in space; the impossibility of evacuating the crew in the event of medical emergencies; limitations in communication between the crew and the Earth; lack of resupply (complete resource autonomy during spaceflight); and the need for improved robotic technology to eliminate human participation in repetitive, dangerous, or mundane housekeeping tasks. These are some of the challenges faced by human space exploration.

The ISS provides an excellent testing ground for space exploration technologies. With frequent resupply and crewmembers spending up to 6 months in orbit, long-duration experiments and comprehensive tests using novel equipment and materials can be completed. Crewmembers are able to troubleshoot and maintain equipment as well as provide direct commentary on the utility of procedures and equipment.

The ISS accommodates scientists as well as engineering and technology developers. Both communities receive unique benefits from using the ISS as a laboratory for their testing and development needs, and are enthusiastic at the prospect of incremental testing and development across the years of ISS utilization.

Commercial Development

The ISS provides an opportunity to test and use commercially developed technology, which is critical to enabling future commercial development and support for human spaceflight. This research seeks to answer the following questions:

• What is the outcome of human space activities applicable to industrial production or to commercial enterprises?

• Will new commercial business be created as a result of space experiments and relevant technology development?

The ISS is a versatile platform for both basic and applied research. By taking advantage of its large space, modern technology, human presence, and microgravity conditions, many kinds of commercial development will lead to new products on Earth.

Early commercial applications from the ISS have included rapid screening of candidate vaccines against microbial illnesses,[20] microcapsules for improved drug delivery to certain types of tumors,[21] and high-quality protein crystals applied to drug design.[22] Other successful commercial activities conducted on the ISS have included high-definition imagery of activities onboard (including popular IMAX films and other content that educates and inspires the public) and visiting travelers launched and returned on Soyuz.

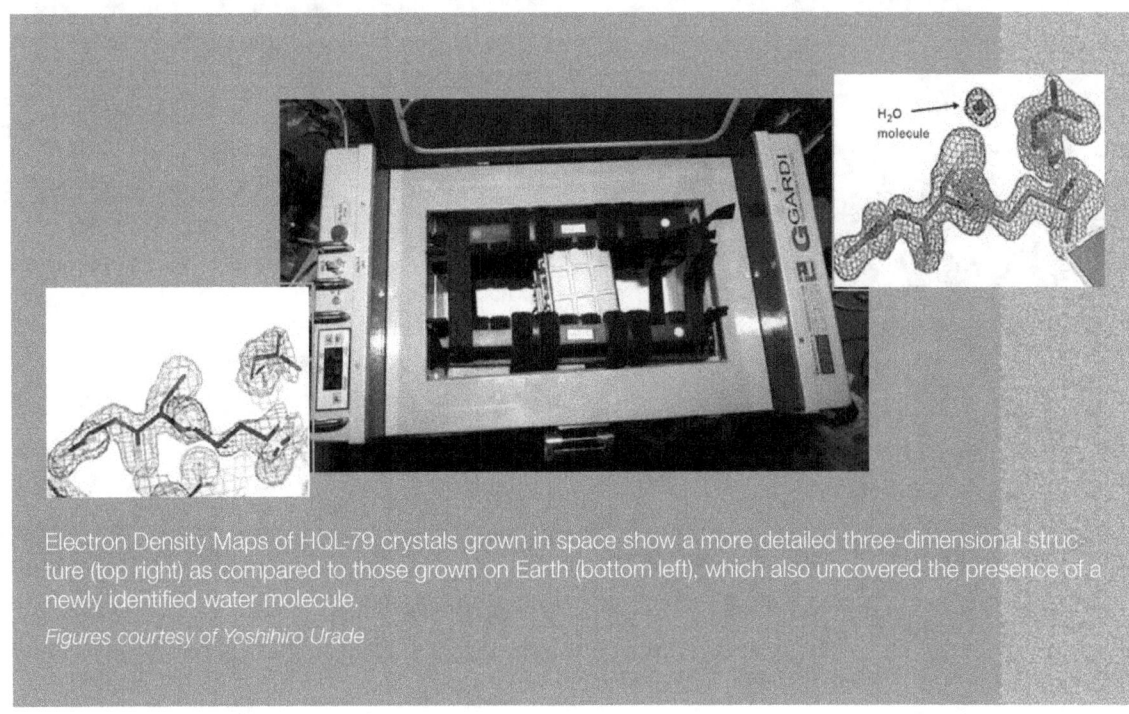

Electron Density Maps of HQL-79 crystals grown in space show a more detailed three-dimensional structure (top right) as compared to those grown on Earth (bottom left), which also uncovered the presence of a newly identified water molecule.

Figures courtesy of Yoshihiro Urade

Education

The engineering and scientific accomplishment of the ISS provides the inspiration and tools to educate students of all ages in science, technology, engineering, and math (STEM). Educational activities on the ISS aim to answer the following questions:

- How can the excitement that students feel about human space activities be used to encourage them to study science, technology, engineering, and mathematics?

- How can the educational activities associated with many ISS experiments be leveraged to reach more students around the world?

- What are the learning outcomes from different types of educational activities linked to the ISS?

Educational activities linked to the ISS offer significant opportunities for educational impacts by inspiring students in their studies of science, engineering, and mathematics through in-flight education downlinks, ham radio contacts from the ISS, and post-flight events. Experiments on the ISS can include student involvement at many different levels—from the development of

Amateur Radio on the International Space Station (ARISS) is used to reach students around the world to inspire interest in math and science. Below, NASA Astronaut Garrett Reisman prepares to use the ARISS system during Expedition 17.

Image: NASA ISS016-E-035837

hardware and training of the crew, to the execution of experiments and ground controls, to the analysis of samples. Experience-based learning on the ISS allows students to develop their own hypotheses and compare results from the ISS to results obtained in their own classrooms. ISS research has involved over 900,000 students in the U.S., and over 31 million more students have participated in educational demonstrations performed by crewmembers onboard the ISS, touching the lives of half the students in the United States.[23]

In Canada, as part of the educational activities related to the first Canadian ISS Expedition crew member, Astronaut Robert Thirsk, 1.85 million students—well over half of the Canadian student population—took part in hands-on learning activities related to ISS science over the course of one school year. Nearly 24,000 students (elementary, intermediate, and post-graduate), either in person or through videoconferencing, participated in five classes from space. The Canadian Space Agency in collaboration with students from the International Space University performed an experiment on optical illusions in space with Thirsk.

To introduce spaceflight to students and the general public, a series of eight podcasts was

Students attending Hanazono Elementary School in Akashi-city, Japan get together for an ARISS contact with NASA Astronaut Sunita Williams in February 2007.

Image courtesy of Satoshi Yasuda, 7M3TJZ

produced from the ISS. This series included: a profile of what it takes to be an explorer; training for an Expedition; and cultural aspects of international space exploration and explanations of on-orbit ISS science. In an effort to encourage healthy living, Thirsk participated in the "Get Fit for Space" challenge along with 35,000 Canadians. Students also learned the challenges of nutritional science—in space and on Earth—by designing grade-appropriate, nutritionally balanced menus for Thirsk based on the challenges of astronaut nutrition in the space environment.

The ISS has already been the focus of numerous educational projects aimed at elementary, secondary, and university students. Long-term educational projects are possible on ISS, thus allowing expansion of the scope of educational activities and making it possible to reach large numbers of students over a period of development.

The High school students United with NASA to Create Hardware (HUNCH) Program provides work experience to inspire middle school and high school students to pursue careers in science and engineering fields. Thirty-one schools in seven states participate in HUNCH, including Alabama, Colorado, Montana, South Dakota, Tennessee, Texas, and Wyoming.

Earth Knowledge Acquired by Middle School Students (EarthKAM), an education activity, allows middle school students to program a digital camera onboard the ISS to photograph a variety of geographical targets for study in the classroom. Photos are made available on the Internet for viewing and study by participating schools around the world. Educators use the images for projects involving Earth science, geography, physics, and social science.

Kids In Micro-g is a student experiment design challenge geared toward grades 5-8. Its purpose is to give students a hands-on opportunity to design an experiment or a simple demonstration that could be performed both in the classroom and aboard the ISS.

A workforce well trained in the disciplines of science, technology, engineering, and mathematics is the foundation for economic success in all industries, including those related to information, energy, and the environment. Conducting educational activities on the ISS and leveraging the educational potential of the research itself have the potential to reach millions of students and ensure the next generation of technology, innovation, and economic development.

Students from Clear Creek High School, Clear Springs High School, Lone Star College, Cypress Woods High School, Splain Middle School, Barbra Jordan High School, and Sterling High School participated in the HUNCH Program during the 2008-2009 school year.

Image: NASA

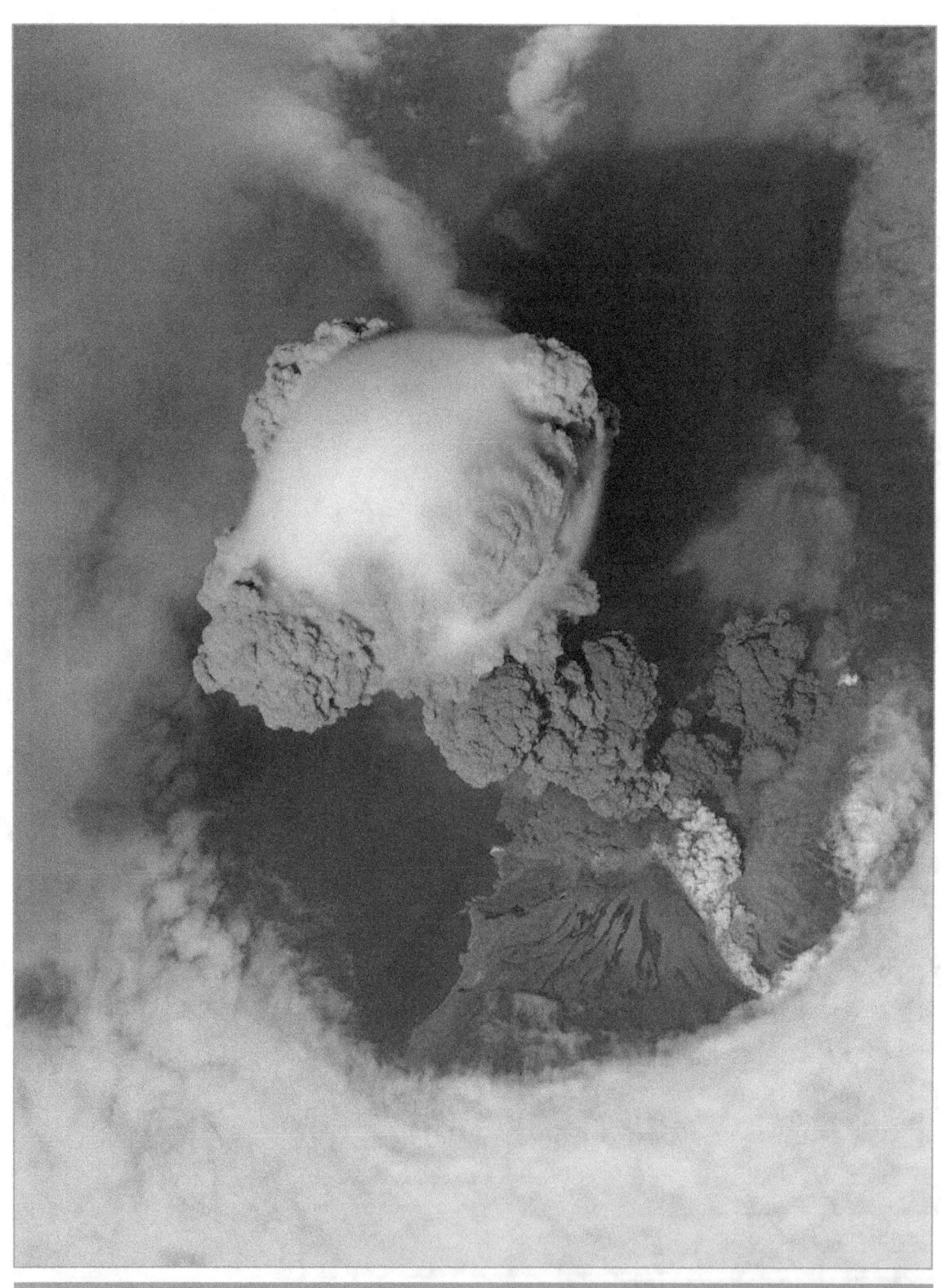

Sarychev Peak Volcano in the early stage of eruption as seen from the ISS on June 12, 2009.
Image: NASA iSS020E009048

The Era of International Space Station Utilization
Space Agency Perspectives

Mark L. Uhran
National Aeronautics and Space Administration

Nicole Buckley and Perry Johnson-Green
Canadian Space Agency

Martin Zell
European Space Agency

Tai Nakamura
Japan Aerospace Exploration Agency

George Karabadzhak and Igor Sorokin
Roscosmos

Julie A. Robinson
National Aeronautics and Space Administration

Summary

Assembly of the ISS is now virtually complete, with all core elements successfully integrated and functionally verified on orbit. Remaining space shuttle flights will pre-position critical systems spares and complete outfitting of research facilities. The ISS Program will transition within 2010 to the full utilization phase. This paper briefly summarizes the intangible benefits brought about through this unprecedented global partnership, and elaborates, at length, on the tangible benefits associated with ISS operations and utilization. The future potential of the ISS is at least as great as the engineering achievements already in hand.

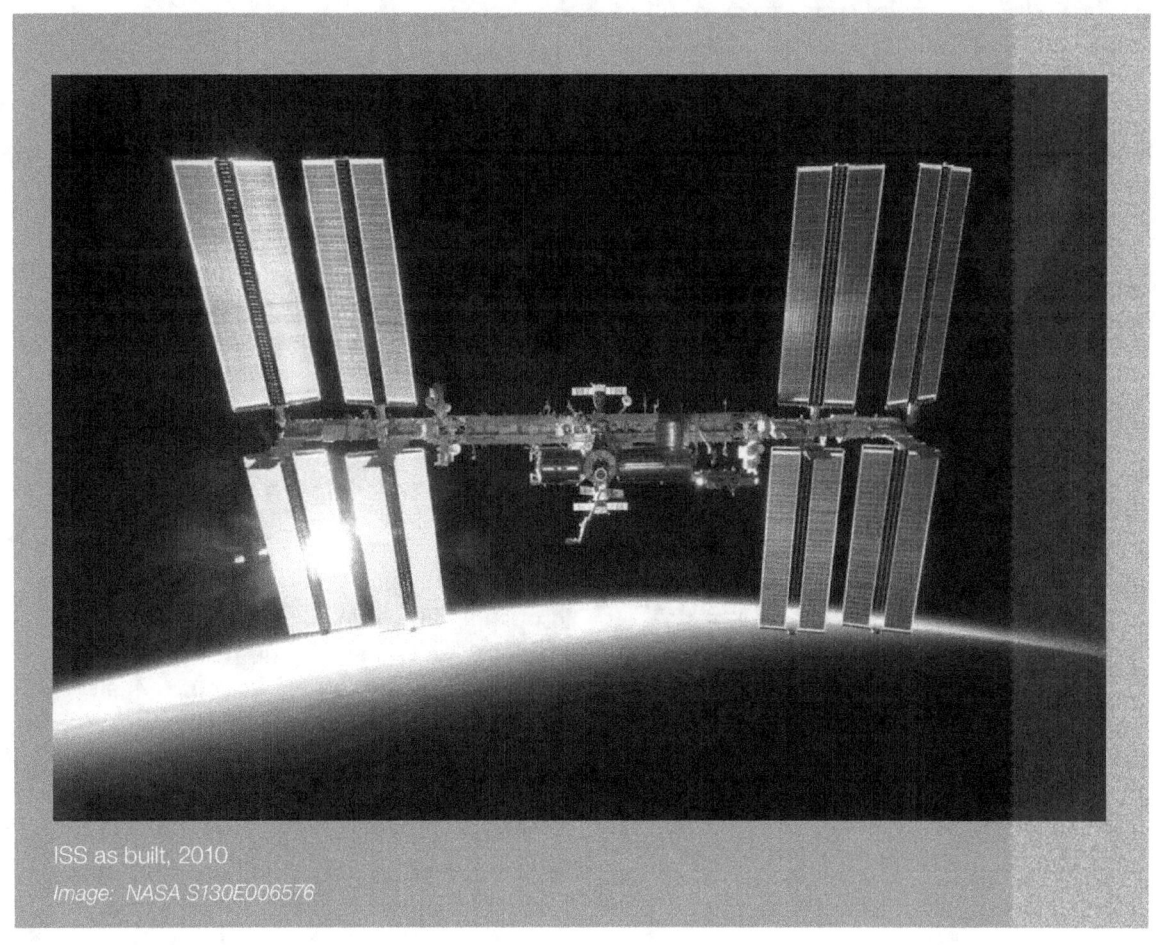

ISS as built, 2010
Image: NASA S130E006576

Introduction

The ISS represents the culmination of over two decades of dedicated effort by an international team of agencies spanning Canada, Europe, Japan, Russia, and the United States. Working in unison on design, development, assembly, and operations in space has set new standards for international partner cooperation and engineering of human-rated space systems. With this leap forward in human space operations come many benefits. The intangible benefits are quick to recognize, but difficult to quantify with precision. Nonetheless, such benefits are real and indeed are at the very basis of the human drive to achieve ever-increasing levels of performance in space.

Alternatively, the tangible benefits are practical, measurable, and unambiguous. The tangibility of the ISS is readily apparent when one compares early concept designs with the physical reality of an approximately 350-metric-ton, permanently crewed, full-service space platform that is now operating with a permanent international crew of six at an altitude of 350 kilometers in a 51.6-degree inclination to the Earth's equator. However, to reap the fullest benefits from this endeavor, one must take the time and effort to reflect on both tangible and intangible benefits and costs. Without this reflection, perhaps the greatest benefit of ISS to future projects and technology will be lost.

Benefits can be viewed retrospectively and prospectively. The benefits of the ISS to date are largely, although not exclusively, in the realm of space systems engineering and operations, while future benefits extend into the vast domains of science and applications. Prior NASA Administrator Michael Griffin captured this aspect succinctly when he remarked,

"It will be the most unique laboratory anyone has ever created. If we use it properly, if funds are appropriated to allow us to use it properly, we will not fail to make groundbreaking discoveries. We do not know what those are, but we know that we will not fail to make such discoveries." [24]

The path of science and applications can change abruptly with the emergence of a transformative technology. The ISS is unquestionably a highly capable scientific laboratory and engineering technology test bed operating in the extraordinarily unique natural environment of space. The future is very promising.

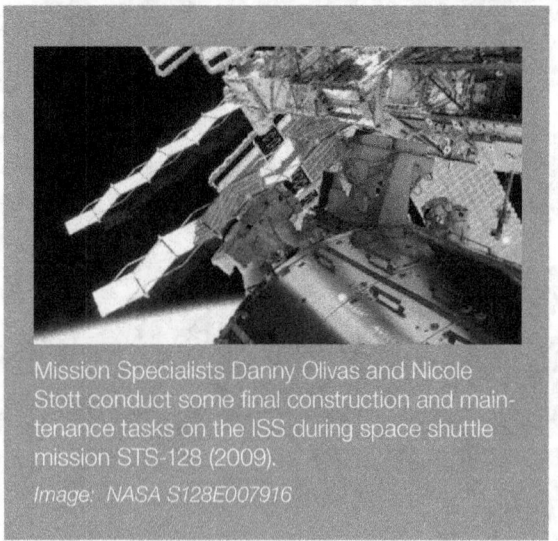

Mission Specialists Danny Olivas and Nicole Stott conduct some final construction and maintenance tasks on the ISS during space shuttle mission STS-128 (2009).

Image: NASA S128E007916

Intangible Benefits

While the intangible benefits are well established and frequently cited, it is useful to quickly review the full scope. These can be generally characterized as: (1) exploring the unknown; (2) human inspiration; (3) international cooperation; (4) global leadership; (5) industrial growth; and (6) educational stimulation. Each has unique features that contribute to a collective benefit that could be attributed to all of space exploration and development, but in this instance is focused on the most recent plateau of achievement—the ISS.

Since the emergence of civilization, exploring the unknown[25] has been a hallmark of progressive

"Spreading out into space will have an even greater effect. It will completely change the future of the human race and maybe determine whether we have any future at all."

- Professor Stephen Hawking

Image: NASA

societies around the world. It's a rich history that spans human and robotic exploration and discovery across half a century. The reward for exploration lies in discovery, and the awareness that each new finding brings us one step closer to understanding our world and reaping the benefits of new knowledge. In the case of ISS, exploration may be the act of assembling, operating, and maintaining a large facility in space, since we are learning what is required to live away from our planet.

Traveling and living in space has consistently evoked public inspiration because it offers hope in a future that involves humankind's evolution outward into our universe. Young and old alike aspire to this adventure and achievement. Travelers to the ISS are among the most sought-after personalities for appearances at events ranging from elementary schools to retirement homes.

The hopes and aspirations of these public audiences are embodied in their desire to live vicariously through the life experiences of space explorers.

The ISS Program has been undertaken by a global alliance that highly values international cooperation. Such an assessment is justified because it recognizes that the efforts of many nations acting peacefully together compound to increase the performance achieved and benefits derived. ISS partners have transcended geopolitical challenges through their cooperative endeavors. Barriers in distance, language, culture, technological maturity, engineering standards, economic competitiveness, industrial capacity, and nationalism have been overcome, thus setting new standards for future international cooperative endeavors in science and technology. The ISS taught us that we must be willing to compromise purely nationalist goals for the greater goals of space exploration. This compromise provided a diversity in capabilities that was not achievable by an individual country.

In terms of technological leadership, the ISS is among the greatest human achievements in history. Global partners have increased their national proficiency in the ability to live and work in the remote and hostile environment of space. This was accomplished through mutual education in an atmosphere of collective problem solving. As a result, leadership has been unequivocally established in large-scale space systems integration—a technical aptitude that simply did not exist prior to the ISS Program. The magnitude of the space and ground systems involved in ISS operations, and distributed across North America, Europe, and Asia, far exceeds that of any prior civil endeavor. The techniques employed in assembly of the ISS, and supported by space vehicles from around the world, have been of an engineering complexity heretofore never imagined. ISS required acceptance of technical standards unique to each of the countries involved and allowed for different approaches from each country.

The completed ISS represents an opportunity for industrial growth through innovation. It will be operated as an "industrial commons," where private firms can access a new environment for R&D of products and services. This is evidenced by two recent initiatives that foreshadow what is to come in the next decade. In one instance, NASA has begun a transition to commercial cargo resupply services that involves procurement of space transportation from entrepreneurial companies that have undertaken private development of vehicles. In another case, a commercially developed water treatment service has been procured, based on recycling of ISS waste carbon dioxide through the Sabatier technique. ISS partner agencies have entered into agreements with private firms that will use the ISS for R&D purposes; and although each agency manages commercial R&D in slightly different ways, public announcements-of-opportunity remain indefinitely open.

Finally, the ISS requires individuals with extensive training in STEM. By their very nature, programs of this magnitude stimulate education as they provide career opportunities for students at the undergraduate, graduate, and post-doctoral levels. In parallel, for primary and secondary school levels, younger students have the chance to participate in human spaceflight through televised broadcasts, experiments, and personal visits by program personnel. To date, over 30 million students have had the opportunity to receive ISS broadcasts. While the number of students stimulated to pursue STEM careers cannot be reliably estimated, it is nonetheless obvious that interest levels are high and widespread.

"…there is significant interest among other Federal agencies in the opportunity to further develop the ISS as an asset for education."

- U.S. Interagency Task Force Report to U.S. Congress

Tangible Benefits

The tangible benefits of the ISS Program are too extensive to address in a summary fashion. In total, over 400 research investigations and 70 educational projects have already been conducted over the past decade.[26] The balance of this paper thus turns to those limited results, as well as to potential future outcomes in two general categories: (1) mission-driven research; and (2) research applied to Earth-based needs. The former includes benefits that apply directly to the primary mission to explore space, while the latter encompasses the need for R&D to advance each partner nation's goals in translating new discoveries into benefits on Earth.

Research to Enable Space Exploration

Human Biomedical Research

The ISS is the best long-duration flight analogue for future human missions involving long transit times. It provides an invaluable laboratory for research with direct application to risks associated with missions beyond LEO. The ISS is being used to identify and quantify risks to human health and performance; identify and validate potential risk mitigation techniques; and develop countermeasures for future missions. The ISS crew is conducting research to develop knowledge in areas of clinical medicine, human physiology, cardiovascular performance, bone and muscle health, neurovestibular medicine, diagnostic instruments and sensors, exercise and pharmacological countermeasures, food and nutrition, immunology and infection, and human behavior.

Astronaut Sunita Williams uses portable testing system for Lab-on-a-Chip Applications Development to assess chemical and biological contaminants to the ISS environment.
Image: NASA ISS015E08353

Engineering Technology Development

The ISS provides a unique opportunity to flight test components and systems in the space environment and optimize subsystem performance. It is the only space-based test bed consisting of pressurized modules and external platforms in open space available for critical systems such as closed-loop life support, extravehicular activity (EVA) suits, energy storage, and automated rendezvous and docking. Characterizing and optimizing long-term system performance in space reduces mission risks and yields next-generation capabilities for long-distance and autonomous vehicle and systems management. As a direct result of the ISS Program, the inventory of space-qualified materials, piece-parts, components, assemblies, subsystems, and systems has expanded rapidly.

Developing robust systems for water and waste recovery, oxygen generation, and environmental monitoring is important as the distance and time away from Earth is extended. The ISS will be used to demonstrate closed-loop life support for oxygen and water systems, microbial detection, air constituents monitoring, and advanced telecommunications.

From 2010 onward, an operational Sabatier system will combine carbon dioxide and excess hydrogen to produce water for the generation of oxygen. A closed-loop life support system can reduce the

Drinkable recycled water from a ground test of the Water Recovery System.
Image: NASA JSC2010E040090

amount of oxygen and water consumables needed by approximately 80 percent. This demonstration is critical for future long-duration human exploration missions. The techniques necessary to maintain these critical systems are also being learned.

Biology Research

The ISS is useful for basic and applied research into understanding the effects of varying gravity levels on cells and organisms, including intracellular activities such as signaling pathways, gene expression, and cytoskeleton structure. The precise mechanism of how cells sense gravity has yet to be determined; by evaluating cells that have experienced the space environment, science comes closer to discovering these mechanisms.

Various experimental studies in this field have been started on the ISS and will lead to better understanding of cellular responses to stress, differentiation of cells into tissues and organs, and how the various systems in the body work together. The potential is great for advances in biotechnological applications such as tissue engineering and regeneration and a decline in the negative effects of aging.

Fluids, Materials, and Processes Research

The ISS is an invaluable experimental platform for research into fluid physics, advanced higher-performance materials, and industrially relevant processes. Initial experiments have shown very interesting results, and the ISS boundary conditions allow the execution of an unprecedented, wide parametric range. In Europe, fluids research in microgravity concentrates on understanding the physics associated with foams, emulsions, colloidal gels, complex plasma, bubbles, boiling devices, and heat exchangers—all of which can improve the design and practical improvement of fluid-based systems on Earth.

Equally important is the research being performed in the area of basic and applied materials. The key objective is to understand the formation of advanced materials, the role that gravity plays in this process, and the accurate measurement of thermo-physical properties.

Since the highly successful Intermetallic Materials Processing in Relation to Earth and Space Solidification (IMPRESS) project, many more applied research projects in both domains are being established with the European Commission in close conjunction with the European Programme for Life and Physical Science (ELIPS) and applications using the ISS. Typically, these projects are characterized by a large, collaborative international team with a strong, multidisciplinary flavor. By making the critical link between experiments in space- and ground-based industrial research for terrestrial applications, the science community and industrial researchers are confident that major technical advances will be made that benefit all our citizens.

Mission Operations Research

Demonstration of human-machine interfaces enable sustained operations over long periods of time. Advances in crew and robotic operations, on-orbit maintenance and repair, and in-space assembly, and demonstrations of crew and cargo transportation vehicles are essential to venture beyond LEO. Assembling six truss segments, eight solar arrays, and four laboratory modules with interconnecting nodes demonstrates the precision and coordination necessary for in-space assembly of large structures.

Autonomous rendezvous and berthing/docking capabilities, essential to complex future space missions, are demonstrated through launch vehicles that transport cargo to the ISS.

Vehicles currently servicing the ISS include the space shuttle, Russian Soyuz and Progress

September 10, 2009, Japanese HIIb launch of HII Transfer Vehicle (left) and March 9, 2008, European Ariane V launch of Automated Transfer Vehicle (right). Both demonstrated new capabilities for automated rendezvous and berthing/docking in space.

Left image: NASA JSC2009E205884
Right image courtesy of ESA

spacecraft, and the European Automated Transfer Vehicle (ATV) and Japanese H-II Transfer Vehicle (HTV). In the future, U.S. Commercial Resupply Service (CRS) vehicles are also anticipated from Space Explorations Technologies Corporation and the Orbital Sciences Corporation.

Robotics plays a critical role in the assembly, maintenance, and resupply of the ISS. The first element of the Canadian Mobile Servicing System (MSS) on orbit was Canadarm2, a 17.6-m (58-ft) robotic manipulator system with 7 degrees-of-freedom. Designed to move payloads as massive as the space shuttle, Canadarm2 can also perform delicate tasks (such as the insertion and extraction of storage platforms from resupply vehicles) using its force-moment sensing capability. Canadarm2 has been essential for the installation of new ISS elements delivered by the space shuttle. It has also provided a stable foothold for spacewalking astronauts during numerous planned and contingency EVAs, allowing them to reach external areas all over the ISS and the space shuttle. The second MSS component, the Mobile Base, allows the MSS to be relocated along the main ISS truss, thereby extending the operational reach of the MSS robotic manipulators. In addition, the

Mobile Base is used for the temporary stowage and relocation of a number of external carriers. The final MSS component is the Special Purpose Dexterous Manipulator, commonly known as Dextre. Equipped with two 3.35-m (11-ft) arms (each with 7 degrees-of-freedom and force-moment sensors), a rotating body, and four tools, Dextre can perform a variety of maintenance tasks normally done during EVAs.

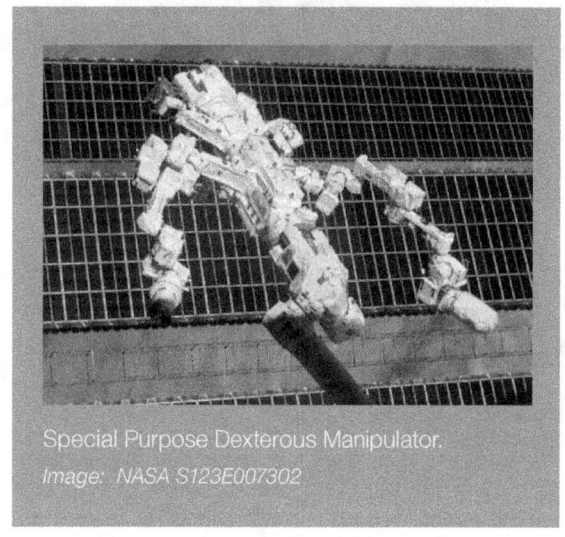

Special Purpose Dexterous Manipulator.
Image: NASA S123E007302

The concept of operation for the MSS has evolved significantly since its arrival on orbit, as past lessons learned and new operational capabilities were incorporated into its software and planning processes. Although initially controlled almost exclusively by astronauts from robotics work stations inside the ISS, Canadarm2 and Dextre's movements are now increasingly controlled from the ground, thereby enabling more efficient use of valuable crew time. When ISS assembly is complete, MSS operations will shift to the capture of resupply vehicles by Canadarm2 and ground-controlled maintenance of the ISS with Dextre.

During the STS-120 mission, while anchored to a foot restraint on the end of the Orbiter Boom Sensor System, Astronaut Scott Parazynski assesses his repair of a torn solar array.
Image: NASA ISS016E009182

Building on the advanced operational knowledge gathered during the ISS maintenance phase, Dextre will also be used to explore novel robotics repair concepts in support of future exploration and on-orbit servicing endeavors.

Development of displays and controls is important for future spacecraft system designs. Training software allows crews to virtually practice spacewalks or robotic tasks before they ever don spacesuits. More than 50 computers control onboard systems, and use some 2.5 million lines of ground-based software code to support 1.5 million lines of software on orbit. Standard communication protocols control crew displays and software tools, while common flight software products, interfaces, and protocols enhance operational practices.

The ISS provides a real-world laboratory for logistics management and in-flight maintenance and repair techniques for future spacecraft. These methods demonstrate an ability to evolve and adapt through daily operations. Common component designs simplify sparing systems, and are used to minimize the number of spares stored on orbit (e.g., common valve design). Interoperable hardware systems include the common berthing mechanism, utility operations

panel, international-standard payload racks, crew equipment, robotic grapple fixtures, and over 500 orbital replacement units that have been designed into distributed systems and elements to reduce the complexity of maintenance operations.

Through thousands of days of operating experience, the ISS is demonstrating the maintainability and reliability of hardware components. Models used to predict this reliability and maintainability are enhanced by measuring the mean-time-between-failure performance on hundreds of components, including pumps, valves, sensors, actuators, solar arrays, and ammonia loops.

ISS crews have had to demonstrate repair capabilities on internal and external systems and components, as well as on hardware not originally designed for on-orbit repair. Such repairs have been performed on malfunctioning spacesuits, computers, treadmill bearings, oxygen generators, carbon dioxide scrubbers, solar arrays, beta gimbals, radiators, and remote power control modules. The flight crews and their ground maintenance counterparts have devised unique solutions that have kept the ISS functioning, including remote maintenance and sustainability procedures and inspection and repair techniques. This experience has helped identify design flaws and redeploy improved systems to orbit.

The ISS provides valuable lessons for current and future engineers and managers—real-world examples of what works and what does not work in space. Developing methods to work with our partners on the ground and in space is critical to providing additional capabilities and solutions to design challenges.

Research Applied to Earth-based Needs

As the ISS is now transitioning from the assembly phase to the full utilization phase, it will be operated as a multilateral space laboratory complex for a broad range of utilization objectives. Each of the ISS partners has a specific utilization strategy for the user community that, however, features a large amount of commonalities and synergies both in research infrastructure and science objectives, which are of mutual benefit to enhance the capabilities and achievements.

At that stage, the benefits will accrue in areas related to Earth-based needs for: (1) improvements in human health; (2) environmental research; and (3) energy systems research. NASA has addressed this expansion of objectives by designating its ISS resources as a U.S. National Laboratory,[27] and other agencies have also worked to ensure ISS access to meet both space-exploration-related and terrestrial needs.

The top researchers on Earth are seeing how the ISS can complement their terrestrial studies. The three basic characteristics of space—variable gravity, exposure to space radiation, and a working confined, extreme environment—are all available with opportunities for real-time troubleshooting and observations, possibilities for follow-up studies, and top-notch facilities.

The ISS is well provisioned for a broad spectrum of science and R&D across the fields of plant and cell biology, astrobiology, human physiology and behavior, chemistry, physics, materials and processes, fluid, and fundamental physics. In the future, ISS research will include disciplines like space physics, Earth observation, and climate change. The international partnership has already invested over $2 billion in research facilities, instruments, and laboratory support equipment, and is prepared to sustain the investment as specific new R&D objectives emerge in the future.[26]

Improvement in Human Health

Early in the ISS assembly period, experiments performed in the Microencapsulation Electrostatic Processing System (MEPS) were performed to improve understanding of fluid mechanics, interfacial behavior, and bio-processing methods for production of encapsulated drug delivery systems. Space-produced microcapsules had properties that improved effectiveness of cancer treatments in a mouse model,[28] and this led to development of a Pulse Flow Microencapsulation System that could replicate the quality on the ground.

NASA licensed the new microencapsulation technology to NuVue Therapeutics, Inc. and others for use in developing ultrasound-enhanced needles and catheters that will be used to deliver microcapsules of antitumor drugs directly to tumor sites. Clinical trials to directly inject microcapsules will begin soon at the M.D. Anderson Cancer Center and the Cancer Center at Mayo Clinic. Other potential applications of the technology include: microencapsulation of

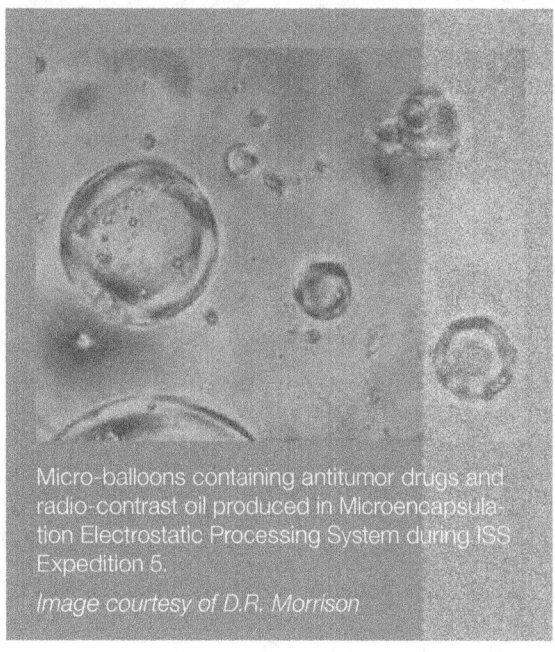

Micro-balloons containing antitumor drugs and radio-contrast oil produced in Microencapsulation Electrostatic Processing System during ISS Expedition 5.

Image courtesy of D.R. Morrison

genetically engineered living cells for injection or transplantation into damaged tissues, enhancement of human tissue repair, and real-time microparticle analysis in flowing sample streams.

As the ISS utilization phase began to ramp up, U.S. companies began pursuing new opportunities in vaccine development. For example, NASA entered into agreements with private firms, such as Astrogenetix, to be pathfinders for the future. Building on results of basic research funded by NASA under prior grants to university investigators,[29] Astrogenetix is now pursing development of vaccines and therapeutic drugs to combat bacterial pathogens. This research is enabled by two phenomena that are unique under microgravity conditions: (1) cells often propagate and exhibit different virulence levels; and (2) genes up- and down-regulate uniquely in the absence of a gravitational force vector.

These phenomena led to discovery of a vaccine target for *Salmonella*-induced food poisoning in 2009, and the company is now seeking investigational new drug status from the U.S. Food and Drug Administration. Follow-on experiments are under way on a variety of bacterial pathogens, including MRSA, which is accountable for almost 20,000 human deaths per year in the U.S. alone.[30]

Improvement in human health is the mission of national institutes around the world. One such example is the U.S. National Institutes of Health (NIH). The NIH entered into a memorandum of understanding with NASA to use the ISS for research.[31] In spring 2009, NIH issued a 3-year rolling announcement for research grants in areas including: (1) cancer; (2) heart, lung, and blood disorders; (3) aging; (4) arthritis and musculoskeletal and skin diseases; (5) biomedical imaging and bioengineering; (6) child health and human development; and (7) neurological disorders and stroke. Research is scheduled to begin by 2011.

A similar agreement was signed with the Agricultural Research Service (ARS) for plant,

Eight syringe mechanisms filled with biological constituents and loaded in a Group Activation Pack are used to test bacterial pathogens for virulence and therapeutic potential.

Image courtesy of BioServe Space Technologies

animal, and environmental research on the ISS.[32] During working sessions with ARS national program leaders in early 2009, specific high-priority research objectives were defined to include: gene transfer efficiency; gene function and biomarker discovery; and viral biodynamics and nutrient bioavailability.

Environmental Research

At the macro level, ISS began serving as an Earth observation platform when the first crew arrived and began using handheld cameras to photograph terrestrial and atmospheric features. At present every month approximately 3,000-4,000 digital images are downloaded. In handheld acquisitions, the spatial resolution that can be achieved allows acquisition of high-quality images and less than 6-m resolution,[33] which have aided researchers in a variety of studies such as urban growth, vegetative cover changes, biogeography, cartography, hydrology, atmospheric research, and the study of aquatic organisms, biomass, coral reefs, endangered species, algal blooms, icebergs, and glacial analysis. Photographs taken over time on ISS provide a record of the changes taking place on Earth and illustrate the potential for remote-sensing instruments that could be developed and tested on ISS.

Hyperspectral cubes represent spatially synchronized sets of spectral data for hundreds of bandwidths, allowing analysis of land and water characteristics.

Image: Goddard Space Flight Center (NASA)

Most recently, the U.S. Naval Research Laboratory has deployed a Hyper-spectral Imager for the Coastal Ocean (HICO) and Remote Atmospheric and Ionospheric Detection System (RAIDS). The objective of HICO is to detect, identify, and quantify littoral and terrestrial geophysical features. Hyper-spectral image data have demonstrated HICO's utility for analysis of land use and land cover, vegetation type, vegetation stress and health, and crop yield. In the ocean, imagery for bathymetry, bottom type, and water optical properties is enabled. These applications are of immediate interest to the U.S. Departments of Agriculture, Commerce, Homeland Security, and the Interior.

The RAIDS sensor package is designed to perform a comprehensive study of upper atmospheric airglow emissions. These observations will be used to develop and test techniques for remote sensing of the neutral atmosphere and ionosphere on a global scale. The package is an array of eight limb-scanning optical instruments covering the wavelength region 550-8,700 Angstroms. The instrument scans the limb of the Earth to measure profiles of airglow from major species in the upper atmosphere. Measurements will be used to determine the composition and temperature of the thermosphere and ionosphere. RAIDS will conduct the most comprehensive survey of the ionosphere and thermosphere in over 20 years.

The Japan Aerospace Exploration Agency (JAXA) recently deployed the Superconducting Submillimeter-wave Limb Emission Sounder (SMILES), which monitors global distribution of trace gases in the stratosphere, including factors related to ozone depletion. The deployment of the Tranquility module with the attached Cupola observation dome will provide additional opportunities for Earth observation experiments.

View of Earth from the newly installed Cupola on the ISS, February 2010.

Image: NASA S130E009953

ESA has recently solicited a Call for Ideas on Climate Change Studies from ISS and received very interesting proposals from highly international research teams, namely in the domains of atmosphere, cryosphere, oceans, and Earth. This constitutes an interesting research capabilities expansion on the ISS. In addition, ESA is also developing the Atmosphere Space Interaction Monitor to study high-altitude optical and gamma-ray emissions associated with large thunderstorms.

Roscosmos has two imaging and spectrometer systems located inside the ISS, *Fialka* and *Rusalka*, operated by cosmonauts. These instruments are used to investigate properties of upper atmosphere and ionosphere basing on measurements of spatial and spectral distributions of the atmospheric species emissions in a wide spectral region allowed by quartz window transparency.

In the environmental sciences, ISS represents among the most sophisticated engineering test beds in the world for oxygen regeneration and water reclamation. A urine processor assembly handles up to 10.5 kg (23.2 lbs.) of condensate, crew urine, and urinal flush water to produce a purified distillate. This distillate is combined with other wastewater sources collected from the crew and cabin and is processed, in turn, by a water processor assembly (WPA) to produce drinking water for the crew. The WPA will process a nominal rate of 22.1 kg (48.8 lbs.) of wastewater per day. A portion of the potable water that is produced is used as feed water to an oxygen generation assembly (OGA). The OGA, in turn, electrolyzes potable water into oxygen and hydrogen byproducts. The oxygen is delivered to the cabin at a selectable rate of 2.3-9.1 kg (5-20 lbs.) of oxygen per day.

The resin used in the microbial check valves in the ISS WPA have now been developed as a commercial water filtration solution by Water Security™ Corporation, and can be used to combat water quality problems anywhere in the world. The commercial system requires little maintenance and no electricity, and provides water that is safe to drink. These systems have been deployed in disaster and humanitarian relief zones in a number of countries including Mexico, Iraq, and Pakistan.

Water filtration plant in Balakot, Pakistan, that was set up following the earthquake disaster in 2005. The unit processes water using gravity fed from a mountain stream.

Image courtesy of the Water Security™ Corporation

Space Sciences

The ISS provides a platform deployment of space science instruments. The JAXA Monitor of All-sky X-ray Image (MAXI) instrument scans the entire sky in X-ray wavelengths during the course of the ISS orbit. Downlinked data are disseminated to research groups via the internet, and alerts are generated for any significant or transient event such as a nova.

AMS-02 will be deployed on the ISS during the STS-134/ULF-6 shuttle flight in 2010. This instrument will detect and characterize high-energy cosmic rays generated by the most energetic events in the universe such as supernovae explosions. This is complementary to studies with large, ground-based particle accelerators. Full characterization of the cosmic ray spectrum is only possible in the space environment, since high-energy particles interact with the Earth's atmosphere. Furthermore, the ISS is an excellent platform for the operation of AMS-02 in that it provides abundant electrical power to operate the instrument. It is expected that AMS-02 will advance our understanding of the universe's origin and particle physics by searching for antimatter, quarks, and particles expected to be associated with dark matter.

Various other major external payloads are deployed or projected for the external sites on the Truss, Columbus, Japanese Experiment Module, and Russian Segment.

Above is an image of the solar cells that provide energy resources to the ISS.

Image: NASA S124E008625

Energy Systems Research

The ISS is a test bed for research on energy-generation, storage, and distribution technologies. The continuing series of Materials on ISS Experiments (MISSE) provides a way to test solar cell materials for accelerated degradation due to exposure to radiation, atomic oxygen, extremes of heat and cold, and other factors. The results will lead to more efficient and durable solar cells for future applications.

In the area of energy storage, ISS currently employs nickel-hydrogen batteries that will wear out and need replacement, so the ISS will convert to use of higher-density lithium-ion (Li^+) batteries. While Li^+ batteries are currently used on the ground at very low energy density levels (e.g., cell phones and calculators), the ISS Program will advance technology by demonstrating Li^+ battery components capable of much greater energy densities for use in electric vehicles of the future.

In terms of power transmission, the ISS represents a suitable platform for the demonstration of microwave, or laser optics, transmission technologies. Space-to-space power relays have obvious applications to future space missions, as well as to ground systems involving power generation that is remote from urban loads.

Conclusion

The benefits afforded by the ISS are both intangible and tangible. The intangibles are well known and quickly recognized. Since the future course of science and applications is impossible to predict, the complete range of tangible benefits will only emerge as scientific and R&D learning progresses, and often many years after the initial discovery. Nonetheless, in the history of science and engineering, new discoveries and subsequent applications have inevitably followed the emergence of disruptive new technologies such as the ISS. As we begin the 21st century, the ISS represents an extraordinary leap forward in civil space technology, and the future potential is at least as great as the engineering achievements already in hand. The ISS is the first step for human exploration of space, and will provide the international partnership an invaluable, permanently accessible and reconfigurable platform with manifold capabilities in space.

Acknowledgements

Completion of the ISS was achieved through the skill and dedication of well over 100,000 government and industry personnel from around the world. While it is impossible to name every contributing individual, it was through the leadership of program managers from Canada, Europe, Japan, Russia, and the United States that the concept ultimately became a reality.

A special acknowledgement to the scientists who have used the ISS during the assembly phase: Your patience has been greatly appreciated, and your enthusiasm for the future is inspiring. To all of the ISS support personnel from each agency: The achievement of assembly and the utilization of the ISS for science is possible because of your hard work and dedication.

Astronaut Patrick Forrester installing MISSE-1 and -2.
Image: NASA STS105-346-007

Biographical Sketches

Manfred Dietel
Charité Berlin, Germany

Manfred Dietel received his diploma in medicine (1973) and was promoted to assistant professor (1980) from the University of Hamburg. In 1983 he became full professor of anatomical and surgical pathology. After becoming Director of the Institute of Pathology, Humboldt-University of Berlin, Charité, he was named Dean of the Medical Faculty Charité, Humboldt-University of Berlin in 1997. He has been Medical Director and Head of the Board of Directors of the Charité from 2001-2004. In 2007, he was named President of the German Society of Pathology. Dr. Dietel is a member of the Editorial Board of the World Health Organization working group on Tumours of the Breast and Female Genital Tract.

Berndt Feuerbacher
International Astronautical Federation, Paris

Berndt Feuerbacher completed his academic education at the Ludwig Maximilian University of Munich and took his Ph.D. in Physics in 1968. Throughout his distinguished career as a scientist studying solid-state physics and material science, he pioneered experimental methods such as photoelectric emission and atom-surface scattering. He initiated the design and construction of a landing probe called "Philae" for the cornerstone ESA Rosetta mission, which presently is on its way to comet Churyumov-Gerasimenko, where it will land in 2014. His scientific results have been published in more than 180 journal papers and in 12 books, and have led to eight patents. In 2006, he was appointed founding director of the new DLR Institute of Space Systems in Bremen. Berndt Feuerbacher was elected president of the International Astronautical Federation 2008-2010.

Vladimir Fortov
Director Joint Institute for High Temperature
Russian Academy of Sciences, Russia

Vladimir E. Fortov received his Ph.D. in Strongly Coupled Plasma Physics in 1971 from the Moscow Institute of Physics and Technology. In 1976, he received his Ph.D. through publication of "Physics of Strongly Coupled Plasma Generated by Intense Shock Waves" by the Russian Academy of Sciences. In 1978, he received a professor's degree in Physics and Chemistry. He is academician of Russian Academy of Sciences, head of the Division of Energetics, Machinery, Mechanics and Control Systems of RAS, and director of Joint Institute for High Temperature of RAS.

David Hart

University of Calgary, Canada
Life Sciences Advisory Committee, Canadian Space Agency

David Hart received his B.A. degree from Northern Michigan University and his Ph.D. in Biochemistry from Michigan State University. In 1983, Dr. Hart moved to the University of Calgary as a Professor of Microbiology & ID and Medicine, as well as more recently the Department of Surgery (2002). He was one of the founding members of the McCaig Centre for Joint Injury and Arthritis Research, the Alberta Bone & Joint Health Institute, and the McCaig Institute for Bone and Joint Health at the University of Calgary. Dr. Hart's research has focused on the molecular and cell biology of wound healing. Dr. Hart has published over 350 original articles, book chapters, and reviews as well as more than 1,100 abstracts.

Charles Kennel

Scripps Institution of Oceanography, USA
Space Studies Board, National Academy of Sciences, USA

Charles F. Kennel was educated in astronomy and astrophysics at Harvard and Princeton. From 1994 to 1996, Kennel was Associate Administrator at NASA and Director of Mission to Planet Earth, the world's largest Earth science program. He became the ninth Director and Dean of the Scripps Institution of Oceanography and Vice Chancellor of Marine Sciences at the University of California, San Diego (UCSD), serving from 1998-2006. He presently is a distinguished professor, emeritus, of atmospheric sciences at Scripps, senior strategist for the UCSD Sustainability Solutions Institute, and co-leads the University of Cambridge/UCSD Global Water Initiative. He has served on many national and international boards and committees: the NASA Advisory Council from 1998-2006 (Chair 2001-2005) and again from 2008 to the present; presently chairs the California Council on Science and Technology and the Space Studies Board of the U.S. National Academy of Sciences; is the 2007 C.P. Snow lecturer at the University of Cambridge; and is a member of the US Review of Human Spaceflight Commission (2009).

Oleg Korablev

Space Research Institute
Russian Academy of Sciences, Russia

Oleg Korablev received a Candidate of Science (Ph.D.) Physics and mathematics; Heliophysics and physics of Solar System and a Doctor of Science (Habilitation) Physics and mathematics; Physics of planets from the Space Research Institute (IKI) in 1992 and 2003, respectively. Since 2002, he is the Deputy Director of Planetary Exploration at IKI. He is involved with a number of space missions, including the Phobos Sample Return Mission. To date, he has 78 refereed publications.

Chiaki Mukai
Space Biomedical Research Office
Japan Aerospace Exploration Agency, Japan

Chiaki Mukai, current Head of the JAXA Space Biomedical Research office, received her doctorate in Medicine in 1977 and a doctorate in physiology in 1988, both from Keio University School of Medicine. She was board certified as a cardiovascular surgeon by Japan Surgical Society in 1989. Dr. Mukai was selected by the National Space Development Agency of Japan (NASDA) in 1985 as one of three Payload Specialist candidates for a U.S. space shuttle mission. During her spaceflight experience on STS-65 (1994) and STS-95 (1998), she logged over 566 hours in space. She has remained a Research Instructor of the Department of Surgery, Baylor College of Medicine, Houston, Texas, since 1992. From 1992 to 1998, she was a visiting associate professor of the Department of Surgery, Keio University School of Medicine, Tokyo; and in 1999 she was promoted to a visiting professor of the university.

Akira Sawaoka
Daido University, Japan

Akira Sawaoka received a Ph.D. in Physics from Hokkaido University in 1968. He was the director of the Research Laboratory of Engineering Materials, Tokyo Institute of Technology, until his retirement in 1999. Since that time, he has been the president of the Daido University. He also has been engaged in promoting applied use of the ISS as a senior counselor of JAXA since 1999. His specialty is international strategies on use of space environment. He was the chairperson of the subdivision on R&D planning and evaluation of the Council for Science and Technology, Ministry of Education, Culture, Sports, Science, and Technology of Japan, and is the program director of key technology of that ministry.

Peter Suedfeld
University of British Columbia, Canada

Peter Suedfeld is Dean Emeritus of Graduate Studies and Professor Emeritus of Psychology at the University of British Columbia. He has conducted field research in space-analogue environments such as Antarctica, laboratory research on stimulus reduction, and archival analyses of the memoirs and diaries of space voyagers. His empirical findings and theoretical propositions concerning human spaceflight and analogue environments have been published in a variety of psychological and medical journals. He has been President of the Canadian Psychological Association, Chair of the Canadian Antarctic Research Program, and Chair of the Life Sciences Advisory Committee of the CSA. He is a member of the Institute of Medicine Committee on Aerospace Medicine and Medicine in Extreme Environments.

Samuel CC Ting

European Organization for Nuclear Research (CERN), Switzerland
Massachusetts Institute of Technology, USA
Nobel Laureate in Physics

Samuel C.C. Ting received his B.S.E. degrees (in Physics and in Mathematics) and his Ph.D. (in Physics) from the University of Michigan. He is the Thomas Dudley Cabot Professor of Physics at the Massachusetts Institute of Technology. Dr. Ting's contributions to the science of physics are numerous. He was awarded the Nobel Prize in 1976 for the discovery of a new kind of matter (the J particle) at the Brookhaven National Laboratory. Currently, he is leading a 16-nation, 600-physicist international collaboration using the ISS and the AMS-02 to probe some of the fundamental questions of modern physics, including the antimatter universe and the origin of cosmic rays and dark matter.

Peter Wolf

Observatoire de Paris, CNRS, LNE, Université Pierre et Marie Curie, France

Peter Wolf received a B.Sc. degree in physics and philosophy from the University of York, G.B., in 1992. He received his Ph. D. from Queen Mary and Westfield College, University of London, in 1997 and his "Habilitation à diriger des recherches" from Université Pierre et Marie Curie in Paris in 2005. He has worked as a physicist in the time section of the Bureau International des Poids et Mesures (BIPM) from 1995 to 2006. Since 2007, he has held a CNRS research position at the Paris Observatory. His research activity is centered on experimental tests of fundamental physics, in particular ground and space tests of gravitation and general relativity and related studies in atomic clocks, atom interferometers, and time/frequency transfer techniques. He is a member of the IAU commission 52 "Relativity in Fundamental Astronomy," of the CNES "Fundamental Physics Advisory Group," and of the ESA "Fundamental Physics Roadmap Advisory Team" and "Physical Sciences Working Group."

Nicole D. Buckley
Director, Life & Physical Sciences
Canadian Space Agency

Christer Fuglesang
Astronaut
Head of Science and Application Division, Human Spaceflight Directorate
European Space Agency

Perry Johnson-Green
Senior Program Scientist, Life & Physical Sciences
Canadian Space Agency

George Karabadzhak
Department head at SUE TsNIIMash. Deputy flight director for RS ISS
Roscosmos

Tai Nakamura
Deputy Director, Space Environment Utilization Center
Japan Aerospace Exploration Agency

Donald Pettit
Astronaut
National Aeronautics and Space Administration

Julie A. Robinson
ISS Program Scientist
National Aeronautics and Space Administration

Igor Sorokin
Deputy Head of Space Stations Utilization Center
Energia

Mark L. Uhran
Assistant Associate Administrator for International Space Station (ISS)
National Aeronautics and Space Administration

Martin Zell
Head of ISS Utilization Department, Human Spaceflight Directorate
European Space Agency

Notes and References

[1] Physical Sciences: Ice crystal grown onboard the ISS for JAXA's Study on Microgravity Effect for Pattern Formation of Dendritic Crystal by a Method of in-situ Observation (Ice Crystal) experiment. *(Image courtesy of JAXA)*

Fundamental Physics: The Alpha Magnetic Spectrometer-02 (AMS-02) hardware schedule to be installed on the ISS in 2010. *(Image courtesy of ESA)*

Life Sciences: *Saccharomyces cerevisiae* (yeast) cells grown on the International Space Station for the Yeast-Group Activation Packs (Yeast-GAP) experiment. *(Image courtesy of Cheryl Nickerson, Arizona State University, Tempe, AZ)*

Human Health: Astronaut T.J. Creamer running on the Combined Operational Load Bearing External Resistance Treadmill (COLBERT). *(Image: NASA ISS022E018811)*

Psychology and Space Exploration: Expedition 9 crew members, Michael Fincke (right) and Gennady Padalka (left), using video and audio channels to communicate with Mission Control Center–Houston on June 18, 2004; in celebration of the recent birth of Fincke's daughter. *(Image: NASA JSC2004E25790)*

Earth and Space Observation: Paris, France, as seen from the ISS in January 2008. *(Image: NASA ISS016-E-21564)*

Exploration and Technology Development: The Synchronized Position Hold, Engage, Reorient, Experimental Satellites (SPHERES) flying in formation onboard the ISS. *(Image: NASA ISS014E17874)*

Commercial Development: The Group Activation Pack–Fluid Processing Apparatus (GAP-FPA) is essentially a microgravity test tube that allows controlled, sequential mixing of two or three fluids in a weightless environment. *(Image courtesy of BioServe Space Technologies, University of Colorado – Boulder, Boulder, CO)*

Education: Astronauts like Canadian Robert Thirsk inspire youth to study science and engineering for future human exploration. *(Image courtesy of CSA and Tomatosphere)*

[2] Roscosmos is scheduled to launch an additional laboratory in 2012 that will be attached to the Russian segment of ISS.

[3] In a related area, foams and emulsions have properties largely governed by the surface tension associated with the interface between the two phases, but gravity intervenes to cause drainage and separation in most cases. One particularly interesting case is that of metal foams, which usually collapse before solidification in the terrestrial environment. Thus, foams of remarkable structure and strength can be demonstrated under low-gravity conditions. Applications range from geophysical flows to bio-engineering (transport of cells or molecules), and many industrial processes are concerned: spray coating, thin film processes (food and pharmaceutical industries), and development of water-repellent surfaces, to name a few.

[4] Studying the thermo- and fluid-dynamics of heat transfer under reduced gravity will provide a unique understanding of the basic processes at stake. Therefore, multi-scale studies are planned on boiling and boiling crisis (bubble nucleation, growth and detachment; bubble/wall interactions), convective boiling (pressure drop and flow pattern prediction), condensation, interfacial heat exchange, and couplings between evaporation and convection. Applications are very diverse: energy conversion (more efficient heat exchangers, boilers, etc.) and cooling of electronics, but also food production and chemical processes.

[5] The Plasma Crystal experiment is an international collaboration whose objective is to create a new physical state of matter—dusty plasma, consisting of electrons, ions, and highly charged (up to 10^5e-) particles—and to study the physical properties of this state. Unlike familiar, highly ionized electron-ion plasma, dusty plasma is suitable for visual observation that makes it a unique physical object providing novel information about: phase transitions, shock waves, turbulence, structure of matter and state equation, and transient properties (viscosity, heat conduction, etc.). Those experiments require continuous research in the microgravity environment.

Annibaldi SV, Ivlev AV, Ratynskaia S, Thomas HM, Morfill GE, Lipaev AM, Molotkov VI, Petrov OF, Fortov VE. Dust-acoustic dispersion relation in three-dimensional complex plasmas under microgravity. 2007. *New Journal of Physics*. 9:327-3335.

Fortov VE, Vaulina OS, Petrov OF, Molotkov VI, Chernyshev AV, Lipaev AM, Morfill G, Thomas H, Rotermell H, Khrapak SA, Semenov YP, Ivanov AI, Krikalev SK, Gidzenko YP. Dynamics of macroparticles in a dusty plasma under microgravity conditions (First experiments onboard the ISS). 2003. *Journal of Experimental and Theoretical Physics*. 96(4): 704-718.

Khrapak S, Samsonov D, Morfill G, Thomas H, Yaroshenko V, Rothermel H, Hagl T, Fortov V, Nefedov A, Molotkov V, Petrov O, Lipaev A, Ivanov A, Baturin Y. Compressional waves in complex (dusty) plasmas under microgravity conditions. 2003. *Physics of Plasmas*. 10(1):1-4.

Nefedov AP, Morfill GE, Fortov VE, Thomas HM, Rothermel H, Hagl T, Ivlev AV, Zuzic M, Klumov BA, Lipaev AM, Molotkov VI, Petrov OF, Gidzenko YP, Krikalev SK, Shepherd W, Ivanov AI, Roth M, Binnenbruck H, Goree JA, Semenov YP. PKE-Nefedov: plasma crystal experiments on the International Space Station. 2003. *New Journal of Physics.* 5:33.1-33.10.

[6] The unique technique of levitation makes it possible to melt and solidify highly reactive liquids in a containerless state. Microgravity allows for accurate measurements of the thermophysical properties of alloys, like thermal conductivity, specific heat, latent heat of fusion, enthalpy, surface tension, viscosity, electrical resistivity, emissivity, melting range, etc. Furthermore, levitation of molten materials allows the creation of novel microstructure selection maps as a function of undercooling and cooling rate. The accurate results obtained from these experiments will be used as input data for sophisticated materials modeling on different length and time scales to optimize industrial metallurgical processes.

[7] Alkemper J, Snyder VA, Akaiwa N, Voorhees PW. The Dynamics of Late-Stage Phase Separation: A Test of Theory. *Physical Review Letters.* 1999; 82:2725.

[8] At present, ACES is the most advanced experiment on atomic quantum sensors in space. ACES is a challenging mission whose aim is to demonstrate the high performances of a new generation of atomic clocks in the microgravity environment of the ISS. The ACES clock signal will be used to generate a stable and accurate time base and to perform precision tests of Einstein's Theory of General Relativity. Besides its scientific relevance, ACES has a key role as pathfinder for follow-on projects aiming at exploiting the high potential of quantum sensors based on cold and ultra-cold atoms. ACES will foster the necessary technological development and, for the first time, will validate in space a series of tools and instruments extremely important for future space missions: from laser and cold atom technology to vacuum techniques, and from space clocks to links for accurate time and frequency dissemination.

[9] Studies of cold atom physics, dark matter, and dark energy now complement the quest for a unification theory that reconciles general relativity and the standard model of particle physics. High-performance quantum sensors such as atomic clocks represent a key technology for accurate frequency measurements and for ultra-precise monitoring of accelerations and rotations. At the same time, studies on ultra-cold atoms and degenerate quantum gases (Bose-Einstein condensates, Fermi gases, and Bose-Fermi mixtures) are rapidly progressing, opening new, exciting perspectives both for fundamental studies and for new atomic quantum sensors based on coherent matter waves.

[10] Early studies of life in extreme environments led to the discovery of heat shock proteins and *Thermus aquaticus* ribonucleic acid (RNA) polymerases important for genetic replication and sequencing studies. Similarly, microgravity presents an extreme environment in which gene expression and biology of unique organisms can be understood and harnessed for benefit on Earth. Brock TD, The Value of Basic Research: Discovery of 117 *Thermus aquaticus* and Other Extreme Thermophiles. 1997. *Genetics,* 146:1207-1210.

[11] Evans CA, Robinson JA, Tate-Brown J, Thumm T, Crespo-Richey J, Baumann D, Rhatigan J. International Space Station Science Research Accomplishments During the Assembly Years: An Analysis of Results from 2000-2008. NASA/TP-2009-213146-Revision A. Also available online at: http://ntrs.nasa.gov/archive/nasa/casi.ntrs.nasa.gov/20090029998_2009030907.pdf.

[12] ESA, European Programme, Columbus Mission Information Kit: http://esamultimedia.esa.int/docs/columbus/infokit/english/11_EuropeanExperimentProgramme_new.pdf.

[13] NASA. Human Research Roadmap. A Risk Reduction Strategy for Human Space Exploration. 2010. http://humanresearchroadmap.nasa.gov/.

[14] WHO - International Agency for Cancer, World Cancer Report, Stewart BW and Kleihues P. (eds.), Lyon 2003

[15] Key studies that should be completed on ISS include: (1) characterization of malignant growth in space and to directly compare theses results with parallel experiments on Earth; and (2) tests of therapeutic approaches through systemic treatment of experimental tumors with different anticancer drugs, including conventional cytostatics, platinum, anthracyclines, targeted drugs (therapeutical antibodies, kinase inhibitor, etc.), and new drugs.

[16] Foale CM, Kaleri AY, Sargsyan AE, Hamilton DR, Melton S, Margin D, Dulchavsky SA. Diagnostic instrumentation aboard ISS: just in time training for non-physician crewmembers. 2005. *Aviation, Space and Environmental Medicine.* 76:594-598.

[17] Mcphee JC (ed) and Charles JC (ed). Human Health and Performance Risks of Space Exploration Missions. May 2009. NASA SP-2009-3405. Also available online at: http://humanresearch.jsc.nasa.gov/files/HRP_EvidenceBook_SSP-2009-3405.pdf.

[18] The external facility locations on the ISS includes Kibo and Columbus modules, Zvezda URM, and the P3 and S3 Trusses. An excellent example of an advanced remote sensing instrument on the ISS is SMILES, a microwave heterodyne spectrometer with superconductive cryogenic detector at 4K, used to study minor species in the Earth stratosphere.

[19] Seurig R, Morfill G, Fortov V, Hofmann P. Complex plasma research on ISS past, present, and future facilities. *Acta Astronautica*. 61(10):940-953.

[20] Other encouraging results in this area were obtained—or have been defined as "promising"—with microbes under development of some new bacterial vaccines (*Salmonella typhimurium,* MRSA, and others). Investment by the company Astrogenetix, Inc. has focused on using the increased virulence of some microbes in microgravity to reduce the time and cost of vaccine development. Arizona State University is independently following other lines of development based on studying microbes in space.

[21] Le Pivert P, Morrison DR, Haddad RS, Renard M, Aller A, Titus K, Doulat J. 2009. Percutaneous Tumor Ablation: Microencapsulated Echo-guided Interstitial Chemotherapy Combined with Cryosurgery Increases Necrosis in Prostate Cancer. *Technology in Cancer Research and Treatment.* 8(3):207-216.

Le Pivert P, Haddad R, Aller A, Titus K, Doulat J, Renard M, Morrison D. 2004. Ultrasound Guided, Combined Cryoablation and Microencapsulated 5-Fluorouracil, Inhibits Growth of Human Prostate Tumors in Xenogenic Mouse Model Assessed by Fluorescence Imaging. *Technology in Cancer Research and Treatment.* 3(2):135-142.

Microcapsules for drug delivery developed through ISS experiments have been used for medical treatment.

[22] For example, high-quality protein crystals made in space have provided detailed data for new drug design (in particular for the development of a novel treatment for Duchenne's muscular dystrophy). Crystal growth in space, a focus of Japanese and Russian ISS utilization, seeks to grow crystals to help make advances in the areas of viral vaccines (AIDS, pneumonia, common cold, influenza) as well as new drugs for Parkinson's, Alzheimer's, and treatment of heart disease. Crystallization of nonbiological materials is also expected to have important commercial applications, including large semiconductor crystals made using a method of molecular beam epitaxy in ultrahigh vacuum and nanomaterials to be used for catalyst development.

Okinaga T, Mohri I, Fujimura H, Imai K, Ono J, Urade Y, Taniike M. 2002. Induction of hematopoietic prostaglandin D synthase in hyalinated necrotic muscle fibers: its implication in grouped necrosis. *Acta Neuropathol.* 104:377–384.

Ohnishi T, Takahashi A, Suzuki H, Omori K, Shimazu T, Ishioka, N. 2009. Expression of p53-regulated genes in cultured mammalian cells after exposure to a space environment. *Biol. Sci. Space.* 23:3-10.

[23] Thomas DA, Robinson JA, Tate J, Thumm T. Inspiring the Next Generation: Student Experiments and Educational Activities on the International Space Station, 2000–2006. NASA/TP-2006-213721. Also available online at: http://ntrs.nasa.gov/archive/nasa/casi.ntrs.nasa.gov/20060015718_2006014780.pdf.

[24] Oral remarks, Congressional signing ceremony for NIH-NASA MOU to Cooperate on Use of the ISS, September 12, 2007.

[25] Logsdon JM, ed., *Exploring the Unknown: Selected Documents in the History of the U.S. Civil Space Program*, Vol. I-VII, NASA SP 4407, 1995-2008.

[26] Harm DL, ed., *Research in Space: Facilities on the International Space Station*, compiled by CSA, ESA, JAXA, Roscosmos, and NASA, 2009.

[27] Uhran ML, Progress Toward Establishing a U.S. National Laboratory on the International Space Station, *Acta Astronautica*, in press, 2009.

[28] Le Pivert P, et al. Ultrasound Guided, Combined Cryoablation and Microencapsulated 5-Fluorouracil, Inhibits Growth of Human Prostate Tumors in Xenogenic Mouse Model Assessed by Fluorescence Imaging. *Technology in Cancer Research and Treatment.* 3(2):135–42.

[29] Wilson J, et al. 2007. Space Flight Alters Bacterial Gene Expression and Virulence and Reveals a Role for Global Regulator Hfq. *Proceedings of the National Academy of Sciences of the United States of America.* 104(41):16299-16304.

[30] U.S. Centers for Disease Control, http://www.cdc.gov/ncidod/dhqp/ar_MRSA.html.

[31] Memorandum of Understanding Between NIH and NASA for Cooperation in Space-related Health Research, 2007.

[32] Memorandum of Understanding Between the USDA ARS and NASA for Cooperation in Space-related Biological and Environmental Research, 2008.

[33] Robinson JA, Evans CA. Space Station Allows Remote Sensing of Earth within Six Meters. *EOS, Transactions of the American Geophysical Union.* 83:185-188, 2002.

To Learn More

Space Station Science
http://www.nasa.gov/iss-science/

Facilities
http://www.nasa.gov/mission_pages/station/
science/experiments/Discipline.html

ISS Interactive Reference Guide
http://www.nasa.gov/externalflash/ISSRG/index.htm

CSA – Canada
http://www.asc-csa.gc.ca/eng/iss/default.asp

ESA – Europe
http://www.esa.int/esaHS/iss.html

JAXA – Japan
http://iss.jaxa.jp/en/

Roscosmos – Russia
http://knts.rsa.ru
http://www.energia.ru/english/index.html

www.ingramcontent.com/pod-product-compliance
Lightning Source LLC
Chambersburg PA
CBHW061009200526
45171CB00009B/532